Prestressed Concrete Design

Prestressed Concrete Design

M.K. Hurst BSc, MSc, DIC, MICE, MI Struct. E

Nanyang Technological Institute, Singapore

London   New York
CHAPMAN AND HALL

First published in 1988 by
Chapman and Hall Ltd
11 New Fetter Lane, London EC4P 4EE
Published in the USA by Chapman and Hall
29 West 35th Street, New York NY 10001
© 1988 M.K. Hurst
Printed in Great Britain by
St Edmundsbury Press Ltd
Bury St Edmunds, Suffolk

ISBN 0 412 28960 1

British Library Cataloguing in Publication Data

Hurst, M.K.
    Prestressed concrete design.
    1. Prestressed concrete construction
    I. Title
    624.1'83412   TA683.9
    ISBN 0 412 28960 1

Library of Congress Cataloging in Publication Data

Hurst, M.K. (Melvin Keith), 1949–
    Prestressed concrete design.

    Bibliography: p.
    Includes index.
    1. Prestressed concrete construction.   I. Title.
    TA683.9.H87   1987   624.1'83412   87-13240
    ISBN 0 412 28960 1

To Jeannie, Eva, Giles and Oliver

CONTENTS

PREFACE

The purpose of this book is to explain the fundamental principles of design for prestressed-concrete structures, and it is intended for both students and practising engineers. Although the emphasis is on design – the problem of providing a structure to fulfil a particular purpose – this can only be achieved if the designer has a sound understanding of the behaviour of prestressed concrete structures. This behaviour is described in some detail, with references to specialist literature for further information where necessary.

Guidance on design of structures must inevitably be related to a code of practice, and the one chosen here is British Standard BS8110: 1985 *Structural Use of Concrete*, which has superseded the earlier British Standard CP110: 1972. These codes relate primarily to the use of concrete in buildings, while concrete bridge structures are covered by British Standard BS5400: 1978, Part 4. The design philosophy and many of the detailed clauses in BS5400 are similar to those in CP110. Many of the clauses in CP110 have been revised in BS8110, although the overall designs of a given structure to the two codes will be similar. No guidance is given in BS8110 for the design of prestressed concrete flat slabs, and reference should be made to Concrete Society Technical Report TR17.

Most of the applications of prestressed concrete in buildings are in the form of simply supported beams, and this is reflected in the many examples throughout the book. Although some of these are examples of bridge decks, the subject of bridge design in general is beyond the scope of this book. Torsion of prestressed concrete in buildings is rarely a problem, and has not been covered here. Information on both of the above topics may be found in the Bibliography. The Bibliography also contains information on other types of prestressed concrete structures such as axial tension and compression members, storage tanks and pressure vessels.

The calculations involved in prestressed concrete design are well suited to implementation by programmable calculator or desk-top microcomputer, with which many design offices are now equipped. However, no examples of programs have been included in the book for reasons of space; the examples in Chapter 13 have been set out in the way they would be performed in a design office.

An overall view of the basic behaviour of prestressed concrete structures is given in Chapter 1. Chapter 2 deals with material properties, while limit state design is outlined in Chapter 3. The detailed considerations in the analysis and design of statically determinate prestressed concrete structures are dealt with in Chapters 4–10. Chapter 11 gives an introduction to statically indeterminate prestressed concrete structures, and Chapter 12 outlines the design principles for the most important applications of such structures in buildings, namely flat slabs. Finally, three full designs of prestressed concrete beams are shown in Chapter 13.

The extracts from British Standards are included by permission of the British Standards Institution, Linford Wood, Milton Keynes, MK146LE, from whom complete copies may be obtained. Extracts from TR17 appear by permission of the Concrete Society, Devon House, 12–15 Dartmouth St, London SW1H 9BL, from whom copies of the complete report may be obtained.

Finally, my thanks are due to Bill Mosley for his initial encouragement in writing this book, to Michael Dunn at Chapman and Hall Ltd, and to Jamillah Sa'adon and Sherlene Lim for patiently typing the original and many revised versions of the manuscript.

SYMBOLS

The symbols used in this book are in accordance with BS8110 wherever possible. They, and other symbols used, are defined below and in the text where they first appear.

A_c	Cross-sectional area of member
$A_{c,slab}$	Cross-sectional area of slab in composite construction
$A_{c,beam}$	Cross-sectional area of beam in composite construction
A_h	Area of shear reinforcement between slab and beam
A_{ps}	Area of prestressing steel
A_s	Area of reinforcing steel
A_s^*	Equivalent area of reinforcement
A_{sv}	Cross-sectional area of two legs of a link
$A\bar{y}$	First moment of area
b	Breadth of section
b_c	Breadth of interface between beam and slab
c	Cover
d	Effective depth
d_r	Drape of tendons
d_n	Depth to centroid of compression block
e	Eccentricity
e_x	Eccentricity in x direction
e_y	Eccentricity in y direction
E_c	Secant modulus of elasticity of concrete
E_{ci}	Modulus of elasticity of concrete at transfer
E_{ct}	Modulus of elasticity of concrete at time t
$E_{c_{eff}}$	Effective modulus of elasticity of concrete
$E_{c,slab}$	Modulus of elasticity of slab concrete
E_s	Modulus of elasticity of steel
f_b	Stress at bottom of section
$f_{b,beam}$	Stress at bottom of composite beam
$f_{b,slab}$	Stress at bottom of composite slab
f_{ci}	Concrete strength at transfer
f_{co}	Concrete stress at level of tendons

f_{cp}	Prestress at level of member centroid
f_{cpx}	Prestress at section x
f_{cu}	Characteristic concrete cube strength
f_{cu_t}	Concrete strength at time t
f_{cv}	Vertical prestress
f_{ht}	Hypothetical concrete tensile stress
f_k	Characteristic strength
f_m	Mean strength
f_{max}	Maximum allowable concrete stress at service load
f'_{max}	Maximum allowable concrete stress at transfer
f_{min}	Minimum allowable concrete stress at service load
f'_{min}	Minimum allowable concrete stress at transfer
Δf_p	Reduction in tendon stress
f_{pi}	Initial stress in tendons
f_{pb}	Tensile stress in tendons at failure
f_{pe}	Effective prestress in tendons after all losses
f_{prt}	Allowable concrete principal tensile stress
f_{ps}	Stress in tendons at service load
f_{pt}	Prestress at level of concrete tensile face
f_{pu}	Characteristic strength of prestressing steel
f_s	Shear stress
f_{st}	Stress in reinforcement
f_t	Concrete stress at top of section
$f_{t,beam}$	Stress at top of beam
$f_{t,slab}$	Stress at top of slab
f_{tu}	Modulus of rupture
f_y	Characteristic strength of reinforcement
f_{yv}	Characteristic strength of shear reinforcement
F_{bst}	Tensile bursting force
F_t	Total tensile force in a section
h	Overall depth of section
h_{eff}	Effective depth of slab at drop panel
I	Second moment of area
I_b	Second moment of area of beam
I_{comp}	Second moment of area of composite section
I_{cr}	Second moment of area of cracked section
I_e	Effective second moment of area
I_g	Second moment of area of gross section
k_{beam}	Beam stiffness
k_{col}	Column stiffness
K	Wobble coefficient
K_t	Transmission length coefficient
L	Span
L_t	Transmission length
L_x	Span in x direction

L_y	Span in y direction
m	Modular ratio
M	Bending moment
M_{cr}	Bending moment to cause cracking
M_d	Dead load bending moment
M_F	Fixed-end moment
M_i	Self weight bending moment
M_0	Bending moment to produce zero stress at tensile face
M_P	Prestress moment
M_{per}	Permanent load bending moment
M_r	Moment of resistance at service load
M_S	Service load bending moment
M_t	Ultimate bending moment transmitted from slab to column
M_x	Bending moment at section x
M_u	Moment of resistance at failure
M'	Moment due to unit point load
p	Loss of prestress force per unit length
P	Prestress force
ΔP_A	Loss of prestress force due to anchorage draw-in
P_e	Effective prestress force after elastic shortening
P_i	Initial force in tendons
P_x	Prestress force in x direction
P_y	Prestress force in y direction
P'	Initial small force in tendons
r	Radius of gyration
r_{AB}	Distribution factor for span AB at joint A
r_{ps}	Radius of curvature of tendons
$1/r$	Curvature
$1/r_b$	Curvature at midspan of beam or support of cantilever
$1/r_c$	Curvature of cracked section
$1/r_u$	Curvature of uncracked section
s_v	Spacing of links
T	Tension
t_{eff}	Effective thickness of section
u	Critical perimeter
v	Vertical shear stress
v_c	Allowable ultimate concrete shear stress
v_h	Horizontal shear stress
V	Shear force
V_c	Ultimate shear resistance of concrete
V_{cr}	Ultimate shear resistance of section cracked in flexure
V_{co}	Ultimate shear resistance of section uncracked in flexure
V_{eff}	Effective shear force in a slab
V_r	Shear resistance of section with nominal reinforcement
V_t	Shear force transferred to a column

w	Distributed load
w_i	Self weight
w_d	Dead load
w_s	Service load
x	Neutral axis depth
x_A	Length of anchorage draw-in effect
x_P	Side of critical perimeter parallel to axis of bending
y	Displacement
y_{P0}	Half the side of loaded area
y_0	Half the side of end-block
$\bar{y}$	Depth to centroid of section
z	Lever arm
Z_b	Section modulus for bottom fibre
$Z_{b,\ beam}$	Section modulus for bottom of beam
$Z_{b,\ comp}$	Section modulus for bottom of composite section
Z_t	Section modulus for top fibre
α	Short-term prestress loss factor
β	Long-term prestress loss factor
β_b	Ratio of bending moments at a section after and before redistribution
γ_f	Partial factor of safety for load
γ_m	Partial factor of safety for materials
δ_a	Long-term deflection under permanent load
δ_{ad}	Anchorage draw-in
δ_b	Short-term deflection under total load
δ_c	Long-term deflection under permanent load
δ_d	Dead-load deflection
δ_e	Expected elongation of tendon
δ_{ex}	Extra elongation of tendon after taking-up of slack
δ_i	Deflection at transfer
δ_M	Deflection of beam due to end-moments
δ_R	Deflection of beam due to point load
ε	Strain
ε_c	Concrete strain
ε_{cu}	Ultimate concrete strain
ε_p	Strain in tendons due to flexure
ε_{pb}	Ultimate strain in tendons
ε_{pe}	Effective prestrain in tendons
ε_{sh}	Shrinkage strain
ε_{st}	Strain in reinforcement
ε_1	Lower yield strain
ε_2	Upper yield strain
η	Degree of prestress
μ	Coefficient of friction
φ	Creep coefficient
σ	Standard deviation

Chapter 1

BASIC PRINCIPLES

1.1 Introduction

Prestressed concrete is the most recent of the major forms of construction to be introduced into structural engineering. Although several patents were taken out in the last century for various prestressing schemes, they were unsuccessful because low-strength steel was used, with the result that long-term effects of creep and shrinkage of the concrete reduced the prestress force so much that any advantage was lost. It was only in the early part of the twentieth century that the French engineer Eugène Freyssinet approached the problem in a systematic way and, using high-strength steel, first applied the technique of prestressing concrete successfully. Since then prestressed concrete has become a well-established method of construction, and the technology is available in most developed, and in many developing, countries.

The idea of prestressing, or preloading, a structure is not new. Barrels were, and still are, made from separate wooden staves, kept in place by metal hoops. These are slightly smaller in diameter than the diameter of the barrel, and are forced into place over the staves, so tightening them together and forming a watertight barrel (Fig. 1.1). Cartwheels were similarly prestressed by passing a heated iron tyre around the wooden rim of the wheel. On cooling, the tyre would contract and be held firmly in place on the rim (Fig. 1.2), thus strengthening the joints between the spokes and the rim by putting them into compression.

The technique of prestressing has several different applications within civil engineering, but by far the most common is in prestressed concrete where a prestress force is applied to a concrete member, and this induces an axial compression that counteracts all, or part of, the tensile stresses set up in the member by applied loading.

Within the field of building structures, most prestressed concrete applications are in the form of simply supported precast floor and roof beams (Fig. 1.3). These are usually factory-made, where the advantages of controlled mass production can be realized. Where large spans are required, *in situ* prestressed concrete beams are sometimes used, and *in situ* prestressed concrete flat slab construction is increasingly being employed. This last technique is often associated with that of the lift slab, whereby whole floor slabs are cast and tensioned at ground level, and then jacked up into their final position.

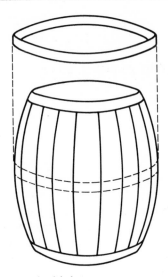

Fig. 1.1 Barrel staves compressed with hoops.

Fig. 1.2 Cartwheel compressed by contracting tyre.

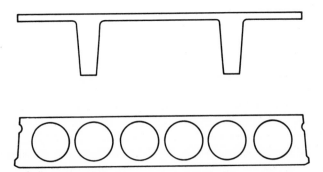

Fig. 1.3 Examples of precast beams.

In the field of bridge engineering, the introduction of prestressed concrete has aided the construction of long-span concrete bridges. These often comprise precast units, lifted into position and then tensioned against the units already in place, the process being continued until the span is complete. For smaller bridges, the use of simply supported precast prestressed concrete beams has proved an economical form of construction, particularly where there is restricted access beneath the bridge for construction. The introduction of ranges of standard beam sections has simplified the design and construction of these bridges (Fig. 1.4).

One of the main advantages of prestressed over reinforced concrete is that, for a given span and loading, a smaller prestressed concrete member is required. This saving of the dead load of the structure is particularly important in long-span structures such as bridges, where the dead load is a large proportion of the total load. As well as a saving in concrete material for members, there is also a saving in foundation costs, and this can be a significant factor in areas of poor foundation material.

Another important advantage of prestressed concrete is that by suitable prestressing the structure can be rendered crack-free, which has important implications for durability and especially for liquid-retaining structures.

A third advantage is that prestressing offers a means of controlling deflections. A prestress force eccentric to the centroid of a member will cause a vertical deflection, usually in the opposite direction to that caused by the applied load. By suitable choice of prestress force, the deflections under applied load can be reduced or eliminated entirely.

Against the advantages listed above must be listed some disadvantages of using prestressed concrete. The fact that most, if not all, of the concrete cross-

Fig. 1.4 Examples of standard bridge beams.

section is in compression under all load conditions means that any inherent problems due to long-term creep movements are increased. From the point of view of construction, a high level of quality control is required, both for material production, and for locating the tendons within the structure.

The technology required for prestressing concrete may not be available in many developing countries, and, if specified, may prove to be uneconomical since all equipment and personnel would have to be imported.

1.2 Methods of prestressing

Methods of prestressing concrete fall into two main categories: *pretensioning* and *post-tensioning*.

(a) Pretensioning

In this method steel *tendons*, in the form of wires or strands, are tensioned between end-anchorages and the concrete members cast around the tendons. Once the concrete has hardened sufficiently, the end-anchorages are released and the prestress force is transferred to the concrete through the bond between the steel and concrete. The protruding ends of the tendons are then cut away to produce the finished concrete member. Pretensioned members usually have a large number of wires or strands to provide the prestress force, since the force in them is developed by bond to the surrounding concrete, and as large as possible an area of surface contact is desirable.

This method is ideally suited to factory production since large anchorages are required to anchor all the tendons, and several members can be cast along the

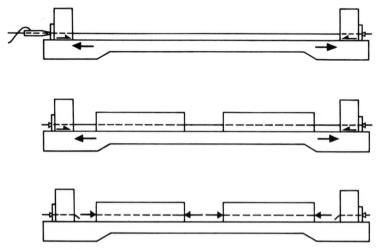

Fig. 1.5 Pretensioning.

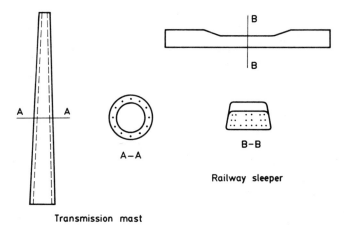

A–A

B–B

Railway sleeper

Transmission mast

Fig. 1.6 Examples of pretensioned members.

same set of tendons (Fig. 1.5). It is important to ensure that the members are free to move along the prestressing bed, otherwise undesirable tensile stresses may be set up in the member when the end-anchorages are released.

Some examples of pretensioned members, other than beams, which are commonly produced are shown in Fig. 1.6.

(b) Post-tensioning

The prestress force is applied in this case by jacking steel tendons against an already-cast concrete member. Nearly all *in situ* prestressing is carried out using this method. The tendons are threaded through ducts cast into the concrete, or in some cases pass outside the concrete section. Once the tendons have been tensioned to their full force, the jacking force is transferred to the concrete through special built-in anchorages. There are various forms of these anchorages and they are considered in detail in Chapter 8. The prestress force in post-tensioned members is usually provided by many individual wires or strands grouped into large tendons and fixed to the same anchorage. The concentrated force applied through the anchorage sets up a complex state of stress within the surrounding concrete and reinforcement is required around the anchorage to prevent the concrete from splitting.

In most post-tensioned concrete applications the space between the tendon and the duct is injected with a cement grout. This not only helps to protect the tendons, but also improves the ultimate strength capacity of the member.

One advantage of post-tensioning over pretensioning is that the tensioning can be carried out in stages, for all tendons in a member, or for some of them. This can be useful where the load is applied in well-defined stages.

An important difference between pretensioned and post-tensioned systems is that it is easy to incorporate curved tendons in the latter. The flexible ducts can

Fig. 1.7 Post-tensioning.

Fig. 1.8 Deflected pretensioning tendons.

be held to a curved shape while the concrete is poured around them (Fig. 1.7). The advantages of having such curved tendons will become apparent later. With pretensioned members, it would be extremely difficult to arrange for a pretensioned curved tendon, although it is possible to have a sharp change of direction, as shown in Fig. 1.8. This involves providing a holding-down force at the point of deflection, and this is another reason why such members are nearly always cast in a factory, or precasting yard, where the holding-down force can more easily be accommodated.

There are other methods available for prestressing concrete (Ramaswamy, 1976), but the ones described above are by far the most common.

1.3 Structural behaviour

Consider a rectangular concrete member with an axial load P applied through its centroid (Fig. 1.9(a)). At any section of the member, the stress at each point in the section is P/A_c, where A_c is the cross-sectional area of the member (Fig. 1.9(b)). The external force is now to be supplied by a tendon passing

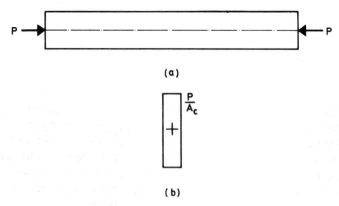

(a)

$\frac{P}{A_c}$

(b)

Fig. 1.9 Axially loaded member.

Fig. 1.10 Axially prestressed member.

through a duct along the member centroid. The tendon is stretched by some external means, such as a hydraulic jack, the stretching force is removed and the force required to keep the tendon in its extended state is transferred to the concrete *via* bearing plates (Fig. 1.10).

As far as the concrete is concerned, the effect is the same as in Fig. 1.9; the stress at each point in a section is P/A_c. (This is not strictly true near the ends of the member where the tendon force is concentrated, but by St Venant's principle it is reasonably true for sections further away from the end.) What has just been described is the situation in a post-tensioned member. In all of the following, the same is also true for a pretensioned member.

If the location of the duct is now moved downwards so that it no longer coincides with the member centroid, but is at a distance e from the centroid (Fig. 1.11(a)) then the stress distribution at any section is no longer uniform. It would be the distribution given by treating the section as though it had an axial force P and a moment Pe acting on it. On the assumption that the member behaviour can be approximated as linearly elastic, the stress distribution can be determined from ordinary bending theory and is shown in Fig. 1.11(b), where Z_t and Z_b are the member section moduli for the top and bottom fibres of the beam respectively.

If the member is now mounted on simple supports at either end and subjected to a uniform load in addition to its own weight (Fig. 1.12(a)) then the stresses at midspan can be determined for a bending moment, M_s which is due to the total uniform load. The resulting stress distribution is shown in Fig. 1.12(b).

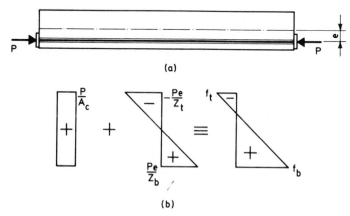

Fig. 1.11 Eccentrically prestressed member.

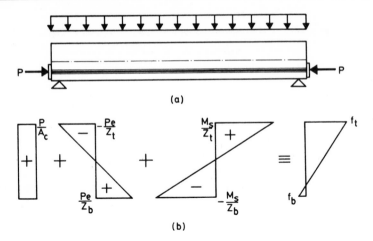

Fig. 1.12 Stresses due to prestress and service load.

Note that the resulting stress distribution has been shown with a net tensile stress at the beam soffit. However, by a suitable adjustment of the values of P and e, this tension can be eliminated and a crack-free member produced. This is the key to the use of prestressed concrete. With reinforced concrete, a certain degree of cracking of the concrete is inevitable. With prestressed concrete it can be eliminated entirely, which has the advantages mentioned earlier.

The use of the terms 'top' and 'bottom' is appropriate with horizontal members such as beams. With vertical members such as masts or tank walls, however, it is more appropriate to consider Z_t and Z_b as the section moduli for the faces of the member with the greater and lesser compressive stresses, respectively, under applied load.

There is another important difference between prestressed and reinforced concrete. With reinforced concrete simply supported beams the minimum load on the beam is of minor importance; it is the maximum load which governs the structural design. However, with prestressed concrete members the minimum load is an important loading condition. Figure 1.13 shows the stresses due to the combination of prestress force and the self weight of the beam. A net tension may occur at the top of the beam, rather than at the soffit as is the case with the

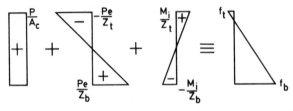

Fig. 1.13 Stresses due to prestress and self weight.

maximum load. This is particularly important since the minimum load condition usually occurs soon after *transfer*, the point when the prestress force is transferred from the tensioning equipment to the concrete and is at its maximum value. Transfer is often carried out soon after the casting of the member (in precast work it is important that transfer is achieved as soon as possible in order to allow rapid re-use of the beds), and the strength of the concrete at this age is usually lower than that when the total, or *service*, load acts on the member.

1.4 Internal equilibrium

If a vertical cut is taken along the beam shown in Fig. 1.11 and the member is separated into free bodies, one containing the steel tendon and one containing the concrete, the forces acting on the free bodies are as shown in Fig. 1.14(a). The respective forces in the concrete and steel are a compressive force P and a tensile force T.

The location of the compressive force in the concrete must, in order to maintain equilibrium, be at the location of the tendon. This may seem obvious for this simple example, but in Section 1.5 a more general treatment is given, where the location of the concrete compression force is not so obvious. Thus, considering the composite member of steel and concrete, internal equilibrium of forces is maintained, and there is no net internal moment, the two forces P and T being equal and opposite, and coincident. This is to be expected from consideration of overall equilibrium, since there is no external axial load on the member.

If, as in Fig. 1.12, the member is mounted on simple supports and a uniform load is applied, there is now a bending moment M_s at midspan, which can be found by considering equilibrium of the beam as a whole. The resultants of the

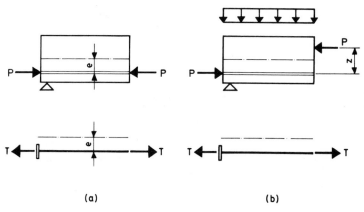

(a)   (b)

Fig. 1.14 Internal equilibrium.

steel and concrete stresses across the midspan section form an internal resisting moment which must balance the bending moment M_s. Since the force in the tendon is fixed in position, defined by the location of the tendon at midspan, it must be the force in the concrete which moves in order to provide an internal resisting couple (Fig. 1.14(b)). The locus of the concrete force along a member is often referred to as the *line of pressure*, a concept which will be useful in dealing with statically indeterminate structures (Chapter 11).

If the section is analysed under the action of a force P acting at a lever arm z from the tendon location, the resulting stress distribution would be that shown in Fig. 1.12(b).

The relationship between prestress force, lever arm and applied bending moment described above is valid right up to the point of collapse of a member and will be used to find the ultimate strength of sections in Chapter 5.

EXAMPLE 1.1 ■■

A simply supported beam with section as shown in Fig. 1.15 spans 15 m and carries a total uniform load, including self weight, of 50 kN/m. If the beam is prestressed with a force of 2000 kN acting at an eccentricity of 400 mm below the centroid, determine the stress distribution at midspan.

Maximum bending moment at midspan $= 50 \times 15^2/8 = 1406.3$ kN m.

Section properties:

$$Z_b = Z_t = 70.73 \times 10^6 \, \text{mm}^3$$
$$A_c = 2.9 \times 10^5 \, \text{mm}^2.$$

Thus the stresses at midspan are:
Top:

$$f_t = \frac{P}{A_c} - \frac{Pe}{Z_t} + \frac{M_s}{Z_t}$$

$$= \frac{2000 \times 10^3}{2.9 \times 10^5} - \frac{2000 \times 10^3 \times 400}{70.73 \times 10^6} + \frac{1406.3 \times 10^6}{70.73 \times 10^6}$$

$$= 6.90 - 11.31 + 19.88$$

$$= 15.47 \, \text{N/mm}^2.$$

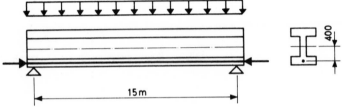

15 m

400

Fig. 1.15

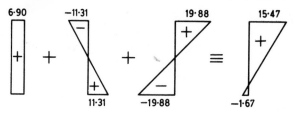

Fig. 1.16 Stresses at midspan of beam in Example 1.1 (N/mm²).

Bottom:

$$f_b = \frac{P}{A_c} + \frac{Pe}{Z_b} - \frac{M_s}{Z_b}$$

$$= 6.90 + 11.31 - 19.88$$

$$= -1.67 \text{ N/mm}^2.$$

The resulting stress distribution is shown in Fig. 1.16.

An alternative approach is to consider the location of the line of pressure in the concrete. From Fig. 1.14(b),

$$z = M_s/P = 1406.3/2000 = 0.703 \text{ m}.$$

Thus the location of the force in the concrete is $(703 - 400) = 303$ mm above the centroid. The stresses may be determined by considering the section under the action of an axial load of 2000 kN and a moment of $2000 \times 0.303 = 606.0$ kN m.

The stresses at midspan are thus:

$$f_t = \frac{2000 \times 10^3}{2.9 \times 10^5} + \frac{606.0 \times 10^6}{70.73 \times 10^6}$$

$$= 6.90 + 8.57$$

$$= 15.47 \text{ N/mm}^2.$$

$$f_b = 6.90 - 8.57$$

$$= -1.67 \text{ N/mm}^2.$$

It is useful to consider what happens at the supports. The bending moment there is zero, and therefore the stress distribution is that shown in Fig. 1.17. A large net tensile stress is produced at the top of the beam. It is thus undesirable to have the same eccentricity at the ends of the member as at the midspan section. This can be overcome by reducing the eccentricity near the supports, as described in the next section. For pretensioned members an alternative is to destroy the bond between the steel and the surrounding concrete by greasing the tendons, or by providing sleeves around them in the form of small tubes, or an extruded plastic coating.

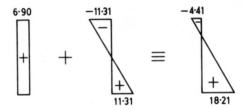

Fig. 1.17 Stresses at the supports of beam in Example 1.1 (N/mm^2).

For sections near to the end of the beam in Example 1.1, the tendon is still in the same location in the section, but the bending moment will be smaller than at the midspan. Thus the value of the lever arm z must change in order to provide a reduced internal resisting moment. The pressure line for a member with no applied load must be coincident with the tendon, in order to satisfy internal equilibrium. Once a load is applied, however, the pressure line must move away from the tendon location in order to provide the internal couple necessary to resist the applied bending moment.

1.5 Deflected tendons

So far all the prestressed concrete members considered have had straight tendons. Consider now a member with a tendon that is deflected at the third-points along its length as shown in Fig. 1.18. The prestress force is no longer horizontal at the ends of the member. The angle θ is usually small, however, and the prestress force may be considered horizontal. The vertical component of the force at the ends of the member is resisted directly by the supports.

In order to determine the stress distribution at midspan, the location of the pressure line is required. To find this, consider free bodies of the concrete and steel respectively (Fig. 1.19) for the left-hand half of the member. The actual direction of the prestress force at the end of the member must now be considered, giving a horizontal component of $P\cos\theta$ and a vertical component of $P\sin\theta$. By considering the free body of the steel tendon, it is clear that there must be a vertical force of $P\sin\theta$ at the deflection point.

The location of the pressure line at the midspan section can be determined by considering equilibrium of the free body of the concrete.

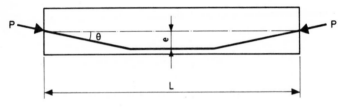

Fig. 1.18 Member with deflected tendons.

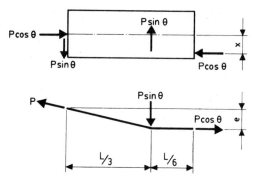

Fig. 1.19 Free bodies of concrete and steel.

Taking moments about the left-hand support:

$$P\sin \theta \ (L/3) = (P\cos \theta)x;$$
$$\therefore x = (L/3) \tan \theta.$$

But

$$\tan \theta = e/(L/3);$$
$$\therefore x = e.$$

The important conclusion is thus that the pressure line in a prestressed concrete member with a deflected tendon, and with no external applied load, is located at the position of the steel tendon for any section along the member. That is, the pressure line is coincident with the tendon profile, as with the case of a straight tendon. Although the above example shows a straight tendon deflected in two places, the same argument applies in the case of a continuously deflected, or draped, tendon as found in most post-tensioned members.

If a cut is made in the beam shown in Fig. 1.18 at a third-point along its length, the free body of the concrete in the left-hand portion will be as shown in Fig. 1.20. Also shown is a shear force V transferred to the left-hand section by the remainder of the beam to the right of the cut. The force in the concrete at the cut is not horizontal and thus there is a vertical component which counteracts the shear force V at the cut section. The shear stresses at that section will therefore be reduced. The determination of the shear resistance of prestressed concrete members is discussed in detail in Chapter 7.

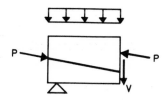

Fig. 1.20 Free body of concrete near a support.

1.6 Integral behaviour

The fact that the pressure line is coincident with the tendon profile for an unloaded member affirms the view of a prestressed concrete member as a single structural element, rather than treating the steel and concrete separately. This aspect is emphasized by considering a vertical concrete member, *prestressed* by a force P through the centroid of the section, Fig. 1.21(a), and comparing it with a similar vertical concrete member loaded with an *external force P* applied through the centroid of the section, Fig. 1.21(b). In the first case, as the force P is increased from zero, there is *no possibility* of the member buckling due to the prestress force alone, whatever the dimensions of the member; failure will eventually occur by crushing of the concrete. In the second case, as the applied force is increased from zero, there *may* come a time when buckling occurs before crushing of the concrete, depending on the dimensions of the member.

The difference between these two examples is important because it illustrates the fundamental behaviour of prestressed concrete. In the first example, if the vertical member is given a small lateral displacement, at any section the pressure line is still coincident with the tendon position, and a uniform stress distribution is obtained. In the second example, if the vertical member is given a small lateral displacement, there is a bending moment induced at any section, and the resulting stress distribution is no longer uniform. In the case of the prestressed vertical member, the effect of the axial force can never be to increase the lateral deflections and so lead to buckling.

Another example is illustrated in Fig. 1.22 where the tendon profile follows the member centroidal axis and the stress at any section along the curved member is always uniform, under prestress force only.

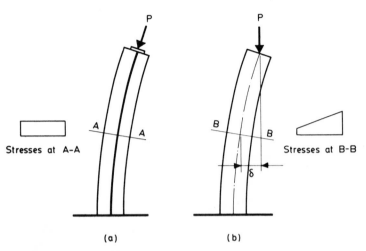

Stresses at A–A

Stresses at B–B

(a)

(b)

Fig. 1.21 Axially loaded and prestressed vertical members.

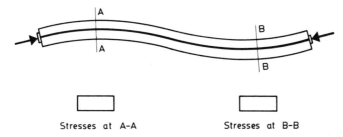

Stresses at A-A   Stresses at B-B

Fig. 1.22 Curved prestressed member.

1.7 Forces exerted by tendons

From Fig. 1.19 it can be seen that, by deflecting a tendon from the straight position, a downwards force is required to maintain the tendon in the deflected position, and this force is transmitted to the concrete as an upwards force. In the case of a continuously curved tendon, there must be a distributed force applied to the concrete to maintain the tendon in position (Fig. 1.23).

In order to determine the value of this force, consider a small, but finite, section of the tendon (Fig. 1.24). If the frictional forces between the tendon and the surrounding concrete are ignored, the force in the tendon at either end of the element Δs is equal to T. If w is the uniformly distributed load on the tendon required to maintain it in position, then, from the triangle of forces,

$$w\Delta s = 2T\sin(\Delta\theta/2).$$

For small changes of angle, $\sin(\Delta\theta/2) = \Delta\theta/2$. If the element is made smaller and smaller, in the limit the force at a point on the tendon is given by

$$w = Td\theta/ds.$$

Now, $d\theta/ds = 1/r_{ps}$, where r_{ps} is the radius of curvature, so that

$$w = T/r_{ps}.$$

Although this force is theoretically directed towards the centre of curvature at any given point, in practice most tendon profiles are reasonably flat and it can be assumed that the force at any point is vertical.

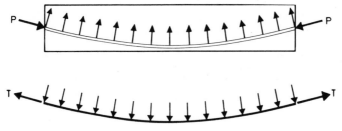

Fig. 1.23 Free bodies of concrete and curved tendon.

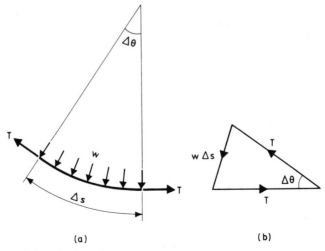

Fig. 1.24 Small length of tendon.

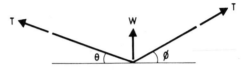

Fig. 1.25 Sharp change of tendon profile.

The vertical force produced by a sharp change of profile, such as that found in pretensioned beams, is shown in Fig. 1.25. In this case,

$$W = T(\sin \theta + \sin \phi)$$

EXAMPLE 1.2

A simply supported beam of length L has a parabolic tendon profile with maximum eccentricity e, as shown in Fig. 1.26. Determine the upwards force on the beam exerted by the tendon and draw the shear force and bending moment diagrams due to the prestress force.

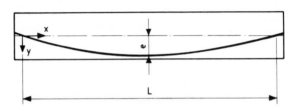

Fig. 1.26

If the parabolic curve is given a set of x and y coordinates with the origin at the left-hand end, the equation of the tendon profile is

$$y = 4ex \, (L - x)/L^2.$$

For a reasonably flat curve, $1/r_{ps}$ may be approximated by d^2y/dx^2,

$$\therefore \ 1/r_{ps} = -8e/L^2;$$

$$\therefore \ w = P/r_{ps} = -8Pe/L^2,$$

where w is an upwards force.
The maximum bending moment in the beam is given by

$$M_{max} = wL^2/8 = (-8Pe/L^2)(L^2/8)$$
$$= -Pe.$$

■ ■

The prestress moment and shear force diagrams are shown in Figs. 1.27(a) and (b) respectively. Note that the prestress moment diagram is a scaled version of the shape of the tendon profile. The moments are negative because the prestress force is below the centroid at midspan, causing a hogging moment in the beam.

The observation that the prestress moment diagram is of the same shape as the tendon profile is true of all statically determinate members. It is particularly useful in drawing the prestress moment diagram, and for determining the deflections (see Chapter 6), for the member shown in Fig. 1.28(a), which has a varying section, but a straight tendon. There can be no vertical loads in this case, since the tendon is straight, but the prestress moment diagram can be drawn simply by considering the distance between the tendon location and the centroid of the member at any section, Fig. 1.28(b).

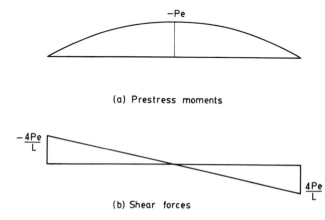

(a) Prestress moments

(b) Shear forces

Fig. 1.27 Prestress moment and shear force diagrams.

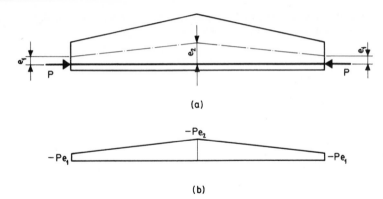

(a)

(b)

Fig. 1.28 Member with varying section.

The fact that a deflected tendon must exert a force on the surrounding concrete is the basis of the 'load balancing' method which has useful application in the design of indeterminate structures, and in particular for analysing prestressed concrete flat slabs (Chapter 12). However, it is not applicable for members with straight tendons, and account must be taken of any moments due to eccentricity at the ends of the member.

1.8 Loss of prestress force

In all the prestressed concrete members considered so far, it has been assumed that the force in the tendon is constant. However, during tensioning of a post-tensioned member, there is friction between the tendon and the sides of the duct. This is caused by changes in curvature of the tendon profile along its length, but even in a straight tendon there is friction present since the tendon does not lie exactly along the centreline of the duct, and there is contact at points along its length.

The effect of friction on the behaviour of post-tensioned members is that, at any section away from the tensioning end, the force in the tendon will be less than that applied to the tendon through the jack. This is shown by considering

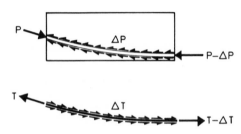

Fig. 1.29 Loss of prestress due to friction.

once again the free bodies of the steel tendon and concrete in a portion of a member (Fig. 1.29).

Friction is only one of the causes of loss of prestress force, and applies to post-tensioned members only. Other causes of loss which apply to both pretensioned and post-tensioned members include initial elastic shortening of the concrete which also shortens the steel tendon, reducing the prestress force. Long-term changes in length of the concrete member due to creep and shrinkage also cause reduction of prestress force. All these effects will be considered in more detail in Chapter 4.

1.9 Degrees of prestressing

When the idea of prestressing concrete was introduced, it was considered that all cracking should be avoided under service load and, further, that the whole section should be in a permanent state of compression. This is often referred to as *full prestressing*. However, at a later stage experiments were carried out using small amounts of tensioned steel to control service load cracking, and larger amounts of untensioned reinforcement, which together with the tensioned steel provided the necessary ultimate strength. This combination of tensioned and untensioned steel is often referred to as *partial prestressing*.

Three classes of prestressed concrete member are categorized in British Standard BS8110. The three classes are summarized in Fig. 1.30 by means of the stress distribution within a section under service load. For Class 1 members, the minimum stress under service load is zero, corresponding to full prestressing. With Class 2 members, some tension is allowed, provided that the tensile strength of the concrete is not exceeded, so that the member is still uncracked. Class 3 members are designed so that cracking occurs, but the extent of the cracking is limited by both tensioned and untensioned steel.

The classification in BS8110 can be seen as a way of relating fully prestressed concrete members to non-prestressed, reinforced concrete members. There have been several parameters proposed to indicate the level of prestressing in a

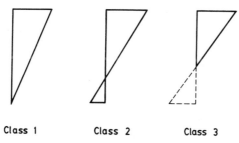

Class 1   Class 2   Class 3

Fig. 1.30 Classes of prestressed concrete member.

member. One such parameter is given by

$$\eta = M_0/M_s,$$

where M_0 is the bending moment at a section to cause zero stress in the concrete at the tensile face, and M_s is the maximum bending moment due to the service load. For a fully prestressed member, $\eta = 1$ and for a reinforced concrete member $\eta = 0$. Class 3 members may be regarded as basically reinforced concrete members with enough prestress introduced to limit service load cracking. The similarity between prestressed concrete and reinforced concrete members is further emphasized by the fact that the behaviour of both types of member is similar at the point of collapse (see Chapter 5).

1.10 Safety

Since high stresses exist in prestressed concrete members at both maximum and minimum load conditions, there must be careful control of the quality of the materials used. In reinforced concrete or steelwork structures these high stresses occur only under maximum load conditions, and are rarely reached. In prestressed concrete structures they are present at all stages of loading. In one sense it can be said that a prestressed concrete structure has been pre-tested, in that the presence of low-standard concrete or steel will generally be detected before the structure enters service.

Since a small change in tendon eccentricity can have a large effect on the stresses induced in a prestressed concrete member, care must be taken during construction that the correct profile for the prestressing steel is maintained during casting of the concrete.

Another important feature of the construction of prestressed concrete structures is the very large jacking forces that are required. Adequate provision must be made to protect site personnel against sudden failure of a steel tendon during tensioning, a not-uncommon occurrence. The large amount of strain energy suddenly released is potentially lethal.

An aspect of prestressed concrete structures which is beginning to concern engineers is how to demolish them. As the early structures reach the ends of their useful lives, the problem arises of how to break up a prestressed concrete member which has such a large amount of energy stored in it. In some cases it is possible to lower the force in the tendons to allow safe demolition. The problem, however, will assume more significance as more prestressed concrete structures require demolition.

Reference

Ramaswamy, G.S. (1976) *Modern Prestressed Concrete Design*, Pitman, London.

Chapter 2

PROPERTIES OF MATERIALS

2.1 Strength of concrete

The strength of concrete is primarily affected by the water/cement ratio, that is the ratio of the weights of mixing water and cement used in a mix. The lower the water/cement ratio the higher the strength, and typical relationships between water/cement ratio and compressive strength at different ages are shown in Fig. 2.1.

A major factor affecting strength is the amount of voids left in the concrete after compaction. The more air contained in the concrete, the more compressible it becomes and the less the strength. It is thus important that the concrete be compacted as fully as possible. It is often the case that the concrete at the top of a horizontally cast member is less well-compacted than at the bottom, leading to lower strength. Another property of concrete affected by poor compaction is the bond developed between the concrete and any steel placed within it. This is

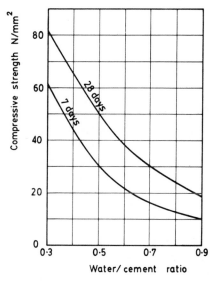

Fig. 2.1 Strength of OPC concrete.

particularly important for pretensioned members, where reliance is made on this bond to transfer the prestress force to the concrete member.

The strength of concrete increases with age, but the rate at which it increases is greatly affected by the curing conditions. Ideally, the concrete should be kept in a moist condition to allow as much hydration of the cement as possible to take place. Most concrete members are cured for the first few days under moist covering and then cured in air.

In the UK the standard compressive strength test is that of a 150 mm cube after 28 days. In the USA, the standard test is on 350 × 150 mm cylinders. The cylinder strength of a given mix is between 70 and 90% of the cube strength. The usual range of concrete cube strengths used in prestressed concrete is 30–60 N/mm², with values at the lower end used for slabs, and those at the upper end for beams. Details of the design of mixes to achieve these strengths are given by the Department of the Environment (1975).

While the tensile strength of concrete is not very important in reinforced concrete design, since its contribution to bending resistance is ignored, in prestressed concrete it is more important to know the tensile strength as well as the compressive strength. An approximate relationship between the modulus of rupture f_{tu} (the theoretical maximum flexural tensile stress determined from a beam test with point loads at the third-points) and the cube strength f_{cu} is given by

$$f_{tu} = 0.59 f_{cu}^{1/2}. \tag{2.1}$$

2.2 Modulus of elasticity of concrete

The modulus of elasticity of concrete is important, not only in estimating deflections of prestressed concrete members but also because some of the losses of prestress force are influenced by the modulus of elasticity. This will be discussed further in Chapter 4.

A typical stress–strain curve for concrete tested in a compression testing machine is shown in Fig. 2.2. The initial portion of the curve is approximately linear, and the modulus of elasticity may be approximated by the slope of the line OA. This is known as the *secant modulus*, and point A is defined at a given concrete stress of $(\frac{1}{3}f_{cu} + 1)$ N/mm², for a rate of load application of 15 N/mm² per minute. This last requirement is important, since the stress–strain curve is dependent on the rate of loading. This is due to the time-dependent deformation of concrete under stress, known as creep (see Section 2.3). An idealized version of the stress–strain curve shown in Fig. 2.2 used for design purposes is shown in Chapter 3.

Values of the secant modulus given in British standard BS8110 for concretes of varying strengths are shown in Table 2.1. They may be used for determining the short-term deflections of prestressed concrete members and the initial losses of prestress force due to elastic shortening. For long-term deflections, the time-

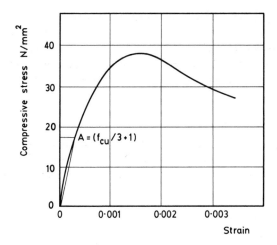

Fig. 2.2 Stress–strain curve for concrete.

Table 2.1 Modulus of elasticity of concrete.

Typical range for the static modulus of elasticity at 28 days of normal-weight concrete

Cube strength (N/mm^2)	Mean value (kN/mm^2)	Typical range (kN/mm^2)
20	24	18–30
25	25	19–31
30	26	20–32
40	28	22–34
50	30	24–36
60	32	26–38

dependent effects of creep and shrinkage should be taken into account (see Chapter 6).

2.3 Creep of concrete

A phenomenon which affects most materials to some extent is that of *creep*, or time-dependent deformation under constant load. Creep is particularly important in concrete, and affects both the long-term deflections and loss of prestress force in prestressed concrete members. The basic mechanism of creep in concrete is that of gradual loss of moisture, causing contraction in the structure of the cement paste in the concrete. The effects of creep in prestressed concrete members are more pronounced than in reinforced concrete members since, in the former, a greater proportion of the cross-section of the member is in compression. A typical curve of creep strain with time is shown in Fig. 2.3. Since creep over a given time interval varies with the level of stress in the concrete, a useful parameter is the *specific creep*, defined as the creep strain per unit stress.

The long-term (30-year) specific creep strain may be determined from the relationship

$$\text{specific creep} = \phi/E_{ct}, \qquad (2.2)$$

where E_{ct} is the modulus of elasticity of the concrete in the long term and ϕ is a creep coefficient determined from Fig. 2.4. There, the effective thickness is defined as twice the cross-sectional area divided by the exposed perimeter. The

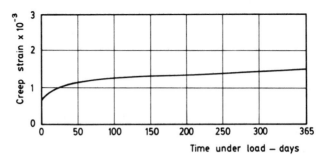

Fig. 2.3 Creep of concrete stored at 20°C with stress/strength ratio 0.7 (Neville, 1977).

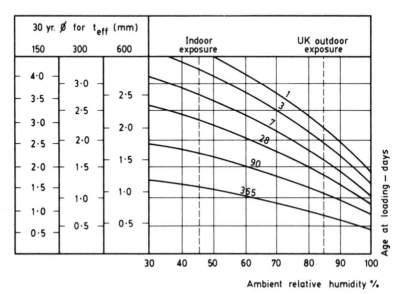

Fig. 2.4 BS8110 creep coefficients.

ambient relative humidity may be taken as 45% and 85% for indoor and outdoor exposure in the UK respectively.

2.4 Shrinkage of concrete

Concrete contains a significant proportion of water, and as the surplus water which has not been used to hydrate the cement evaporates, the concrete member

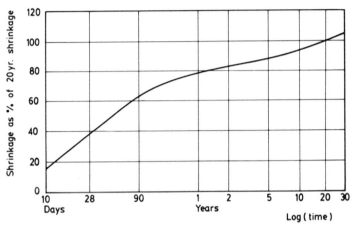

Fig. 2.5 Average shrinkage of concretes stored at 50–70% RH (Neville, 1977).

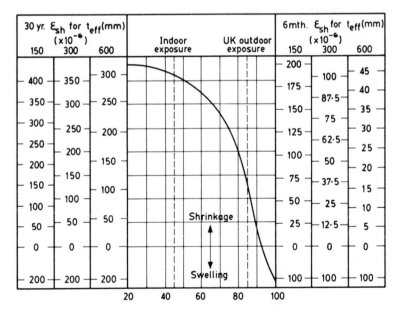

Fig. 2.6 BS8110 shrinkage strains.

will shrink. The amount of shrinkage is dependent on the environmental conditions surrounding the concrete, and is independent of the external load on the member. If the concrete is in a dry windy climate, the loss of moisture will be much greater than if the concrete is kept in a moist condition.

Shrinkage of concrete varies with time, and a typical relationship is shown in Fig. 2.5. The initial rate of shrinkage decreases, and by the end of one year approximately 80% of the long-term shrinkage has taken place.

The values of short- and long-term shrinkage strain to be used for design purposes may be found from Fig. 2.6. This relates to concretes with a water content of approximately 190 litres/m^3. Shrinkage may be regarded as proportional to water content in the range 150–230 litres/m^3.

The effects of creep and shrinkage on loss of prestress force and long-term deflections will be discussed in Chapters 4 and 6 respectively. Further information on both these effects may be found in Neville (1977).

2.5 Lightweight concrete

Several notable prestressed concrete structures have been built in recent years with concretes made from lightweight aggregates. Apart from the primary advantage of a saving in weight, these concretes also offer better fire resistance and insulation properties.

The aggregates used may be naturally occurring, such as pumice, or manufactured, such as expanded blast furnace slag. The density of the concretes produced with these aggregates is in the range 1300–2000 kg/m^3, compared with 2400 kg/m^3 for normal-density concretes.

For lightweight concretes the strength is usually dependent on the strength of the aggregates. The 28-day cube strengths obtainable range from 17 N/mm^2, using expanded clay aggregate, to 60 N/mm^2, using pulverized-fuel-ash aggregate. The modulus of elasticity of lightweight concrete is between 50 and 70% of that for normal-density concrete. Shrinkage and creep effects are usually greater, and in the absence of more detailed information from the aggregate supplier, the shrinkage strain may be taken as $400–600 \times 10^{-6}$, and the specific creep as $0.7–0.9 \times 10^{-4}$ per N/mm^2 (Abeles and Bardhan-Roy, 1981).

Further design information on the use of lightweight aggregate concrete may be found in Part 2 of BS8110.

2.6 Steel for prestressing

There are several different types of steel used for prestressing, covered by respective British Standards: (i) wire, to BS5896: 1980; (ii) strand, to BS5896: 1980; (iii) bars, to BS4486: 1980. Wires vary in diameter from 3 to 7 mm, and have a carbon content of 0.70–0.85%. The wires are drawn from hot-rolled rods, which have been subsequently heated to 1000°C and then cooled to make them suitable for drawing. After several drawing operations which reduce the diameter of the wire and increase the tensile strength, the wires are wound on to capstans with diameter 0.6–0.7 m. This is known as the *as-drawn* condition, and the steel is supplied in mill coils, suitable for pretensioning.

However, as-drawn wires will not pay out straight from the coils, but they can be pre-straightened, to make them suitable for threading through post-tensioning ducts, by heating them for a short time, or heating them while subjected to high tension. Both these processes also increase the elastic range of the wires over the as-drawn condition. The former type of wire is known as *stress-relieved* wire, and the latter as *stabilized* wire. Stress-relieved wire is also termed *normal-relaxation* wire, while stabilized wire is also known as *low-relaxation* wire, since its relaxation properties (see Section 2.7) are much better than for either as-drawn or stress-relieved wire.

For pretensioned concrete members, the prestress force is transferred to the concrete by bond between the steel and concrete. This bond is substantially increased if indentations are made on the wire surface, and also if the wire is *crimped*, that is given an undulating, instead of a straight, shape. These two processes are shown in Fig. 2.7. The crimp pitch varies between five and twelve times the wire diameter, and the wave height varies between 5–12% of the diameter for helical crimping and between 10–25% of the diameter for uniplanar crimping.

Strand is produced by spinning several individual wires around a central core wire. Modern strands comprise seven wires with overall diameters ranging from 8.0–18.0 mm. A strand can be spun from as-drawn wires to produce an *as-spun* strand, or it can be heat-treated after spinning to produce either *normal-* or *low-relaxation* strand.

In order to make the strands, six wires are helically wound round a central straight wire (Fig. 2.8(a)). They can be produced as either *standard* or *super* and can also be drawn through a die to compact them, when they are known as *drawn* strands; the cross-section is then as shown in Fig. 2.8(b). All three types of strand can be produced with either normal- or low-relaxation properties.

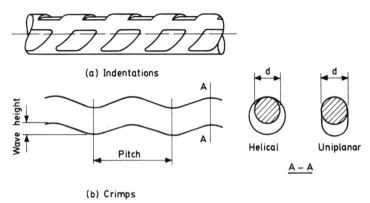

Fig. 2.7 Wire for prestressing.

Tables 2.2 Properties of prestressing steel.

BS	Type of tendon	Nominal diameter and steel area		Nominal tensile strength f_{pu}		Specified characteristic load (kN)		Maximum relaxation (%) after 1000 h	
						Breaking load (A)	0.1% proof load or load at 1% elongation	at 70% of A	at 80% of A
		mm	(mm²)		N/mm²		0.1% proof load		
5896	Cold-drawn steel wire (pre-straightened)	7	(38.5)	1670		64.3	53.4	Class 1	
		7		1570		60.4	50.1	8	12
		6	(28.3)	1770		50.1	41.6	Class 2	
		6		1670		47.3	39.3	2.5	4.5
		5	(19.6)	1770		34.7	28.8		
		5		1670		32.7	27.2		
		4.5	(15.9)	1620		25.8	21.4		
		4	(12.6)	1770		22.3	18.5		
		4		1670		21.0	17.5		
							Load at 1% elongation		
	Wire in mill coils (As-drawn wire)	5	(19.6)	1770	N/mm²	34.7	27.8	For all wires	
		5		1670		32.7	26.2		
		5		1570		30.8	24.6		
		4.5	(15.9)	1620		25.8	20.6	10	
		4	(12.6)	1770		22.3	17.8		
		4		1720		21.7	17.4		
		4		1670		21.0	16.8		
		3	(7.1)	1860		13.1	10.5		
		3		1770		12.5	10.0		

BS	Type	Diameter (mm)	(Area mm²)	N/mm²	kN	0.1% proof load		
5896	*7-wire steel strand* standard	15.2	(139)	1670	232	197	Class 1 8	12
		12.5	(93)	1770	164	139	Class 2 2.5	4.5
		11.0	(71)	1770	125	106		
		9.3	(52)	1770	92	78		
	Super	15.7	(150)	1770	265	225		
		12.9	(100)	1860	186	158		
		11.3	(75)	1860	139	118		
		9.6	(55)	1860	102	87		
		8.0	(38)	1860	70	59		
	Drawn	18.0	(223)	1700	380	323		
		15.2	(165)	1820	300	255		
		12.7	(112)	1860	209	178		
4486	*Hot-rolled or hot-rolled and processed alloy steel bars* Hot rolled	40	(1257)	1030	1300	1050	3.5	6.0
		40	(804)		830	670		
		25	(491)		505	410		
		20	(314)		325	260		
	Hot rolled and processed	32	(804)	1230	990	870	3.5	6.0
		25	(491)		600	530		
		20	(314)		385	340		

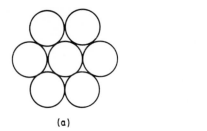

(a)   (b)

Fig. 2.8 Prestressing strands.

Hot-rolled alloy-steel bars vary in diameter from 20 to 40 mm, and are stretched once they have cooled in order to improve their mechanical properties. They may be ribbed, to provide a continuous thread, or smooth with threads at the ends of the bars. In both cases the threads are used to anchor the bars or to provide a coupling between adjacent bars.

The properties of the various types of wire, strand and bar are summarized in Table 2.2. Each manufacturer will have his own range of products, and reference should be made to trade literature before deciding on the choice of prestressing steel.

An important point to consider with all the types of steel described above is that their high strength is produced by essentially a cold-working process. Thus, during storage and construction care must be taken not to expose the steel to heat, from causes such as welding. Further information on the manufacture and properties of prestressing steel may be found in Bannister (1968).

2.7 Relaxation of steel

Relaxation of steel stress is similar to creep in concrete in that it is time-dependent deformation under constant load. The amount of relaxation depends on time, temperature and level of stress. The standard test for relaxation determines the value after 1000 hours at 20°C and it is these values which are shown in Table 2.2. Typical variations of relaxation with time are shown in Fig. 2.9, at 20°C and at an initial stress level of $0.7f_{pu}$, where f_{pu} is the breaking strength.

For a 1000-h relaxation value of 1% at 0.7 f_{pu} and 20°C, this would typically increase to 6.5% at 100°C under constant initial stress, and to 1.5% for an initial stress of 0.75 f_{pu} at constant temperature.

Two classes of relaxation are specified in British Standard BS5896, Class 1 corresponding to stress-relieved, or normal-relaxation, wires, and Class 2 corresponding to stabilized, or low-relaxation, wires.

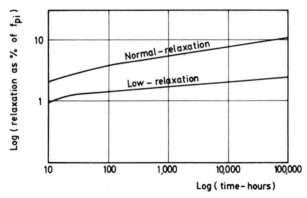

Fig. 2.9 Relaxation of steel at 20°C with initial stress $0.7f_{pu}$ (Courtesy British Ropes Ltd).

2.8 Stress–strain curves for steel

Typical stress–strain curves for prestressing steel are shown in Fig. 2.10. These high-strength steels do not possess the same well-defined yield point as mild steel, and so the *proof* stress is defined as that stress for which, when the load is removed, there is a given permanent deformation. The deformation specified in

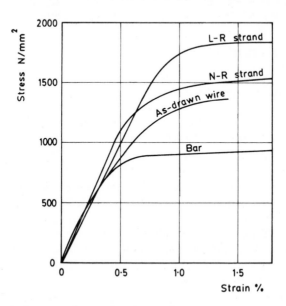

Fig. 2.10 Stress–strain curves for prestressing steel.

Table 2.3 Modulus of elasticity of steel.

Type of steel	$E_s(kN/mm^2)$
Wire to BS5896 Section 2	205
Strand to BS5896 Section 3	195
Rolled and stretched bars to BS4486	206
Rolled, stretched and tempered bars to BS4486	165

the British Standards for prestressing steel is 0.1% elongation, and so in Table 2.2 the minimum or 0.1% proof loads are given. Alternatively, the load to cause 1% elongation may be specified. Also specified in Table 2.2 is the characteristic strength of the steel. This term will be explained in more detail in Chapter 3, but for the present it can be regarded as the breaking load of the wire or strand.

The heat-treating processes applied to the as-drawn wire not only decrease the relaxation of the steel, but also increase the proof stress, thus extending the linear elastic range.

The moduli of elasticity E_s given in BS8110 for the different types of steel are summarized in Table 2.3. The lower value of E_s for strand compared with wire is due to the slight twisting action of strand under tension, giving a greater axial extension than a wire of similar cross-sectional area. Also given in BS8110 are idealized stress–strain curves for use in determining the steel stresses in a prestressed concrete member; these are shown in Chapter 3.

2.9 Corrosion of steel

As with steel reinforcing bars, prestressing steel must be protected from attack by moisture permeating the surrounding concrete. In pretensioned members this is prevented by having adequate cover to the tendons and also by using concrete with a sufficiently low water/cement ratio, which is usually the case for the high-strength concretes used for prestressed concrete. The ducts in post-tensioned members are usually grouted after tensioning, and this process also provides a bond between the tendons and the surrounding concrete. Particular attention must be paid to tendons which are to be left exposed. A combination of greasing and coating with plastic has been used successfully.

Another form of corrosion which can affect wires and strands is known as *stress corrosion*, and arises from a breakdown of the structure of the steel itself. Small cracks appear and the steel becomes brittle. Little is known about stress corrosion, except that it occurs at high levels of stress, to which prestressing steels are continually subjected.

Further information on corrosion and its prevention may be found in Libby (1971).

References

Abeles, P. W. and Bardhan-Roy, B. K. (1981) *Prestressed Concrete Designer's Handbook*, Viewpoint, Slough.

Bannister, J. L. (1968) *Concrete*, August, pp. 333–342.

Department of the Environment (1975) *Design of Normal Concrete Mixes*, Her Majesty's Stationery Office, London.

Libby, J. A. (1971) *Modern Prestressed Concrete*, Van Nostrand Reinhold, New York.

Neville, A. M. (1977) *Properties of Concrete*, Pitman, London.

Chapter 3

LIMIT STATE DESIGN

3.1 Introduction

The early codes of practice for reinforced concrete and steelwork design were based on the concept of working stresses. That is, the maximum elastic stresses in the materials under the working, or service, loads were compared with allowable values, based on the stresses at failure of the material divided by a suitable factor of safety. It was soon realized that concrete is an inelastic material, although elastic methods are still suitable for service load conditions for prestressed concrete members. This was reflected in the first code of practice for prestressed concrete, British Standard CP115, which adopted separate approaches for the service load and ultimate load behaviour of members. This was an early example of *limit state design* – identifying all the possible loading conditions for a member and choosing the most critical as the basis of design, while checking the other load conditions afterwards. A general approach based on limit state principles, which also identifies factors affecting the performance of structures other than loading, was incorporated into British Standard CP110, covering both reinforced and prestressed concrete, and it has been retained in BS8110.

3.2 Limit states

The limit state concept involves identification of the various factors that affect the suitability of a structure to fulfil the purpose for which it is designed. Each of these factors is termed a limit state, and if any of them is not satisfied, then the structure is deemed to have 'failed'. The consequences of failure, however, vary considerably between the limit states, and this may be accounted for by using different factors of safety for each.

The two principal limit states for most structures are:

(a) Ultimate limit state

The most important of the ultimate limit states is:

(i) *Strength.* The structure must be able to withstand, with an acceptable factor of safety against collapse, the loads likely to act upon it. Collapse can occur in several ways, including fracture of an individual member, instability of the structure as a whole, or by buckling of part of the structure. An adequate factor of safety against collapse under accidental overloading must also be provided, although this is generally lower than that provided for the service loads.

Other ultimate limit states which must be considered are:

(ii) Fire resistance. The structure must stand for sufficient time to allow any occupants to escape. The fire resistance of a concrete structure is mainly determined by the concrete cover to the steel, since it is the steel strength which is greatly reduced with increasing temperature.

(iii) Fatigue. For structures subject to cyclic loading, this could be important, especially in the case of prestressed concrete structures, where the stress level in the prestressing steel is very high.

(b) Serviceability limit state

There are several serviceability limit states, the two most important of which are:

(i) Deflection. The deflections of the structure under the service load must not be excessive, otherwise damage to finishes, partitions or cladding may result.

(ii) Cracking. Excessive cracking may not only be unsightly, but may lead to excessive ingress of water into the concrete, leading to corrosion of the steel.

Other limit states which may need to be considered include:

(iii) Durability. If the concrete is too permeable then the risk of corrosion of the steel is increased. Possible attack by an aggressive environment, such as seawater, must also be considered. The main factors influencing the durability of prestressed concrete structures are the concrete mix proportions, the cover to the steel, and grouting of post-tensioning ducts.

(iv) Vibration. This is important in such structures as machine foundations.

For any given structure, the limit states relevant to the loads acting upon, and the environment surrounding, the structure must be identified and the most critical defined.

For most prestressed concrete structures the design process entails initially considering the serviceability limit state of cracking, and then checking the ultimate strength limit state. Reinforced concrete design, by contrast, is usually based on the ultimate strength limit state, with later checks on the serviceability limit states.

3.3 Characteristic loads and strengths

An important concept in limit state design is that of a *characteristic* load, or material strength. This is a concept taken from the theory of probability, and reflects the fact that, for instance, the concrete strength to be used in calculations should not be the mean strength determined from a series of cube tests, but rather a figure much lower than the mean, such that there is an acceptable probability that any given test result will be less than a specified value. This value is known as the *characteristic* strength, f_k.

The spread of cube test results from any mix of the same nominal proportions will generally follow a normal distribution curve, as shown in Fig. 3.1. The value of f_k is defined such that the shaded area corresponding to cube results below f_k is 5% of the total area under the curve. In terms of probability, this is equivalent to saying that there is a 1 in 20 chance of any test result falling below f_k, which is considered an acceptable probability.

From the mathematical properties of the normal distribution curve, it can be shown that

$$f_k = f_m - 1.64\sigma,$$

where σ is the standard deviation of the set of test results.

The distribution of strength for both prestressing and reinforcing steel also follows a normal curve, and in all subsequent work f_{cu} will be taken as the characteristic concrete cube strength, f_y as the characteristic yield strength of untensioned steel, and f_{pu} the characteristic breaking strength of prestressing steel.

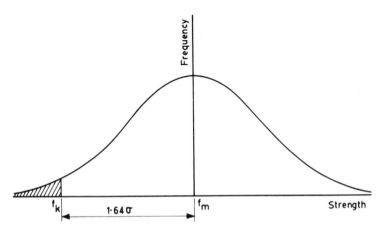

Fig. 3.1 Concrete strength distribution.

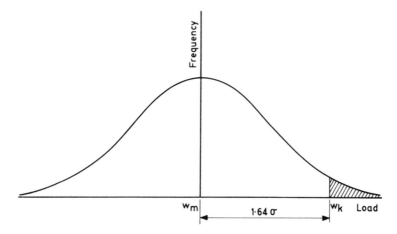

Fig. 3.2 Load distribution.

In the case of loading on a structure, although there are at present insufficient data available fully to justify the use of probability methods, the limit state approach in BS8110 assumes that the distribution of load on a structure also follows a normal curve, as shown in Fig. 3.2. The characteristic load on the structure, w_k, is defined as that load for which there is a 1 in 20 chance of its being exceeded. In terms of the normal distribution curve this means that the shaded area is 5% of the total area under the curve, and the value of w_k is given by

$$w_k = w_m + 1.64\sigma.$$

3.4 Partial factors of safety

The characteristic values of strength and load are those which are used in calculations rather than average values, but for several reasons they are adjusted by the use of partial factors of safety.

For material properties, the characteristic strengths are divided by a partial factor of safety γ_m, so that the design strength is given by

$$\text{design strength} = f_k/\gamma_m$$

The use of γ_m is to account for various factors which are difficult to quantify individually; it is found that the use of an overall factor to cover them is satisfactory. The value of γ_m is chosen to take account of the variability of the strength properties of the material used, the difference between site and laboratory strengths, the accuracy of the methods used to determine the strength of sections, and variations in member geometry which affect this strength. The value of γ_m for concrete is higher than that for steel to allow for the lesser degree of control which can be exerted over the production of concrete compared with that of steel.

Table 3.1 Partial factors of safety for loads.

	Dead		Imposed		Earth and water pressure	Wind
	Adverse	*Beneficial*	*Adverse*	*Beneficial*		
1. Dead and imposed (and earth and water pressure)	1.4	1.0	1.6	0	1.4	—
2. Dead and wind (and earth and water pressure)	1.4	1.0	—	—	1.4	1.4
3. Dead and wind and imposed (and earth and water pressure)	1.2	1.2	1.2	1.2	1.2	1.2

Table 3.2 Partial factors of safety for materials.

Reinforcement	1.15
Concrete in flexure or axial load	1.50
Shear strength without shear reinforcement	1.25
Bond strength	1.4
Others (e.g. bearing stress)	$\geqslant 1.5$

For loading, the characteristic load is multiplied by a partial factor of safety γ_f, so that the design load is given by

design load $= w_k \times \gamma_f$.

The values of γ_f accommodate the inherent uncertainties in the applied loading, the analytical methods used to obtain the bending moment and shear force distributions within a structure, and the effects on the design calculations of construction tolerances. The value of γ_f also reflects the importance of a given limit state, and the consequences of its being exceeded. Thus the highest values of γ_f are assigned to the ultimate limit states.

An advantage of using partial factors of safety is that their values may be amended if more or less information is available. Thus if particularly good quality control of concrete production under factory conditions can be demonstrated, then a lower value of γ_m may be used. Conversely, if concrete production is known to be in the hands of unsupervised, unskilled labour, then γ_m may be increased. Some loads can be predicted more accurately than others, such as dead loads, or loads due to soils and liquids, and so a lower value of γ_f may be justified than when the loading is more difficult to predict.

The values of γ_f and γ_m recommended in BS8110 are given in Tables 3.1 and 3.2 respectively. Although not specifically stated in BS8110, γ_m for both concrete and steel at the serviceability limit state is 1.0, except for tensile stresses in prestressed concrete, where γ_m is 1.3.

For the first two load combinations in Table 3.1, the adverse γ_f is applied to any loads producing a critical condition for a given section, while the beneficial γ_f is applied to any load which lessens this condition.

Consideration should be given to the effects of large accidental overloads caused, for instance, by an explosion. In this case the loading should be the dead load plus one-third of the imposed load (or the full imposed load, if this is permanent) and one-third of the wind load, with γ_f taken as 1.05. The values of γ_m for concrete and steel should be 1.3 and 1.0 respectively.

3.5 Stress–strain curves

The short-term stress–strain curves for concrete and prestressing steel in Chapter 2 are idealized in BS8110 for design purposes, and these are shown in

Figs. 3.3 and 3.4 respectively, where the stresses have been divided by the appropriate γ_m. Note that the factor of 0.67 in Fig. 3.3 relates the cube strength of concrete and the flexural strength in an actual member.

These curves are suitable for determining the strength and short-term deformations of members, but for long-term deformations the modulus of elasticity of concrete should be modified, as described in Chapter 6.

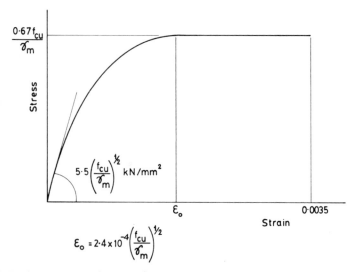

Fig. 3.3 Design stress–strain curve for concrete.

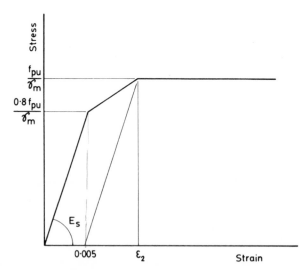

Fig. 3.4 Design stress–strain curve for prestressing steel.

3.6 Loading cases

It was noted in Chapter 1 that, unlike reinforced concrete, for prestressed concrete members the minimum load condition is always an important one. Most prestressed concrete members are simply supported beams, and so the minimum bending moment at any section is that which occurs immediately after transfer of the prestress force, M_i. This is usually due to the self weight of the member, although in some cases additional dead load due to finishes may be present. Prestressed concrete members are often moved from their formwork soon after transfer and it can be assumed that the prestress force is a maximum, since only the short-term losses have occurred. The concrete is usually weaker than it is under the service load and so the allowable stresses at transfer are less than at the service load.

An important consideration in composite construction, where a precast beam acts together with an *in situ* slab is the bending moment, M_d, due to the dead load, that is the weight of the beam and slab. In this case most of the losses have usually occurred and so the prestress force should be taken as a minimum.

The maximum bending moment, M_s, occurs at the service load and at this stage most of the long-term losses have occurred and the prestress force is a minimum.

For continuous beams, consideration must also be given to pattern loading, with the adverse and beneficial values of γ_f given in Table 3.1 applied to the various spans to give the maximum and minimum bending moments for each span. Thus a third combination must be considered, that of minimum bending moment and minimum prestress force.

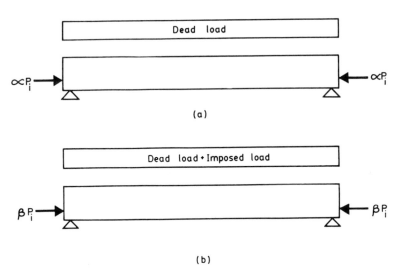

Fig. 3.5 Principal load cases.

The principal loading cases for a simply supported member are summarized in Fig. 3.5, where P_i is the initial prestress force and αP_i and βP_i are the effective prestress forces for the respective loading cases (see Chapter 4).

An important loading case for many prestressed concrete members is that which occurs when a slender member is being lifted, either for storage in a casting yard or factory, or into its final position on site. Lateral bending may occur in slender members, and a realistic estimate of the imperfections induced during the casting of the member must be made. Some guidance on estimation of the lateral bending moments induced is given in Rowe *et al.* (1987).

3.7 Allowable stresses

When considering the serviceability limit state of cracking of prestressed concrete members, three classifications of structural members are given in BS8110, as described in Chapter 1:

CLASS 1: no flexural tensile stresses;

CLASS 2: flexural tensile stresses, but no visible cracking;

CLASS 3: flexural tensile stresses, but surface crack widths not exceeding a maximum value (0.1 mm for members in aggressive environments and 0.2 mm for all other members).

The normally accepted width of cracks in reinforced concrete is 0.3 mm, and it is argued in a report of the Concrete Society (1983) that this should also be adopted as the maximum crack width for Class 3 members not subjected to an aggressive environment. This would have the effect of allowing more economical designs.

The allowable compressive and tensile stresses for bonded Class 1 and 2 members at transfer and under service load are shown in Table 3.3, where f_{ci} and f_{cu} are the characteristic concrete strengths at transfer and service load respectively. The higher proportion of cube strength allowed at transfer is due to the fact that the transfer condition is temporary, and an overstress is considered acceptable. For continuous members in bending, the maximum compressive stress at service load may be increased to $0.4 f_{cu}$ near to the supports since the

Table 3.3 Allowable stresses for Class 1 and 2 members.

	Transfer	Service load
Compression	$0.50 f_{ci}$	$0.33 f_{cu}$
Tension:		
Class 1	$1.0 \, \text{N/mm}^2$	0
Class 2: Pretensioned	$0.45 \, f_{ci}^{1/2}$	$0.45 \, f_{cu}^{1/2}$
Post-tensioned	$0.36 \, f_{ci}^{1/2}$	$0.36 \, f_{cu}^{1/2}$

vertical restraint of the support induces a biaxial state of stress in this region. The maximum compressive stress for axially loaded members if $0.25 f_{cu}$.

The allowable tensile stress for pretensioned members has been derived by applying a γ_m of 1.3 to the modulus of rupture given by Equation 2.1, while that for post-tensioned members is limited further since tests have shown that cracks propagate earlier around post-tensioning ducts than around pretensioning tendons.

Where the full service load is of a temporary nature and is high compared with the normal service load, the tensile stresses in Table 3.3 may be increased by up to $1.7 \, N/mm^2$, provided that under the normal service load the stress is compressive, thus ensuring that any cracks which do open up close again. In this case pretensioned members should have tendons well distributed throughout the tensile region, and untensioned reinforcement should be placed in post-tensioned members. These measures limit the extent of any cracking and should also be adopted for the tensile zones at transfer.

No distinction is made in BS8110 between allowable tensile stresses for members with bonded and unbonded tendons. However, for unbonded members, it is recommended (Clark, 1982) that the same guidelines should be followed as for flat slabs (see Chapter 12), namely that in sagging moment regions the tensile stress be limited to $0.15 f_{ci}^{1/2}$ or $0.15 f_{cu}^{1/2}$ at transfer and service load respectively, and in hogging moment regions the stress be limited to zero for both load cases.

The allowable stresses in Class 3 members are considered in Chapter 5.

3.8 Fire resistance

The fire resistance period for a structure, or a portion of a structure, is defined as that period for which the structure must remain intact during a fire, in order that all occupants may escape. The requirements for a given structure are contained in the regulations applicable to that structure.

The fire resistance of prestressed concrete members, as with reinforced concrete members, is governed by the loss of strength of the steel with increase in temperature, rather than by loss of concrete strength. Generally, failure of prestressed concrete members is only likely at temperatures above 400°C. The high-strength prestressing steels lose a greater proportion of their strength at a given temperature than do reinforcing steels, being approximately one-half of the characteristic strength at 400°C for strands. Thus greater fire resistance, in the form of cover to the steel, is required in prestressed concrete members than in reinforced concrete members. The cover required is usually greater than that required for protection against corrosion, and so should be considered at an early stage in the design process. For good fire resistance of all concrete members, attention must be paid to detailing, since reinforcement is required near the member faces to prevent spalling. Class 3 members can withstand very high temperatures better than Class 1 or 2 members, since their greater

Table 3.4 Concrete cover for fire resistance.

| Fire resistance | Nominal cover | | | |
| | Beams | | Floors | |
	Simply supported	Continuous	Simply supported	Continuous
Hours	mm	mm	mm	mm
0.5	20	20	20	20
1	20	20	20	20
1.5	20	20	25	20
2	40	30	35	25
3	60	40	45	35
4	70	50	55	45

proportion of lower-strength steel is less affected by high temperatures. Lightweight aggregate concretes exhibit better fire resistance than normal-density concretes, since less spalling occurs and better insulation is afforded to the steel.

The nominal covers specified in BS8110 for varying periods of fire resistance and type of structural element are shown in Table 3.4. The lesser cover values given for continuous members in all types of prestressed concrete, compared with the simply supported condition, are due to the fact that continuous members have the ability to redistribute the load if one region loses strength in a fire.

Detailed information is given in Part 2 of BS8110 on determining the fire resistance period of given structural elements, and also specified in the code are minimum overall dimensions of concrete members to provide given fire resistance periods. Further information on fire resistance of prestressed concrete members may be found in Abeles and Bardhan-Roy (1981).

3.9 Fatigue

For prestressed concrete members subjected to repeated loading, the fatigue strength must be considered. The major areas where fatigue failure could occur are in the concrete in compression, the bond between the steel and concrete, and the prestressing steel.

The compressive stress level in concrete above which failure could occur is approximately $0.5 f_{cu}$, and in the life of most prestressed concrete members this ensures that failure due to fatigue in the concrete is unlikely. Bond failures have been observed in tests on short members such as railway sleepers, but for most applications this does not present a problem.

Although the stress levels in prestressing tendons are high, the range of stress in the tendons is usually small, since the imposed load is not a large proportion of the total service load. Fatigue failure of tendons has been observed in tests, generally associated with high concentrations of stress in the vicinity of cracks in the concrete. If the concrete remains uncracked, the range of stress in the steel is small. Class 1 members thus exhibit much better fatigue resistance than Class 3 members.

The fatigue strength of prestressing tendons may be taken to be between 65% and 75% of the characteristic strength for two million load cycles, and this is usually much greater than the maximum stress in the steel under service load.

One area which has been identified as potentially troublesome is in pretensioned members where the tendons have been deflected. There are stress concentrations at the deflection points and deflected tendons should be avoided if the members are to be subjected to cyclic loading.

The stress variations in unbonded tendons are transferred to the anchorages rather than distributed to the surrounding concrete as with bonded tendons. Unbonded tendons should thus generally be avoided if fatigue is a consideration.

A method of determining the fatigue resistance of prestressed concrete members may be found in Warner and Faulkes (1979), and general information on fatigue found in Abeles and Bardhan Roy (1981).

3.10 Durability

There have been many failures of structures in recent years which can be attributed to poor durability of concrete. These failures have generally not resulted in actual collapse of a structure, but serious corrosion of reinforcement

Table 3.5 Exposure conditions.

Environment	Exposure conditions
Mild	Concrete surfaces protected against weather or aggressive conditions
Moderate	Concrete surfaces sheltered from severe rain or freezing whilst wet
	Concrete subject to condensation
	Concrete surfaces continuously under water
	Concrete in contact with non-aggressive soil
Severe	Concrete surfaces exposed to severe rain, alternate wetting and drying or occasional freezing or severe condensation
Very severe	Concrete surfaces exposed to sea water spray, de-icing salts (directly or indirectly), corrosive fumes or severe freezing conditions whilst wet
Extreme	Concrete surfaces exposed to abrasive action, e.g. sea water carrying solids or flowing water with pH $\leqslant 4.5$ or machinery or vehicles

Table 3.6 Concrete cover for durability.

Conditions of exposure	Nominal cover (mm)				
Mild	25	20	20	20	20
Moderate	—	35	30	25	20
Severe	—	—	40	30	25
Very severe	—	—	50	40	30
Extreme	—	—	—	60	50
Maximum free water/ cement ratio	0.65	0.60	0.55	0.50	0.45
Minimum cement content (kg/m^3)	275	300	325	350	400
Lowest grade of concrete	30	35	40	45	50

has sometimes occurred, significantly weakening the structure. The clauses in BS8110 relating to the durability of concrete have been amended from those in the former CP110 in an effort to provide more durable structures.

Five degrees of exposure of concrete members are identified in BS8110, and these are shown in Table 3.5. The minimum cover requirements for all types of steel in prestressed concrete members are shown in Table 3.6, which also gives maximum water/cement ratios and minimum cement contents to be used. The cover requirements relate to normal-density concrete with 20 mm maximum aggregate size. Where the prestressing steel passes through a duct, the minimum cover to the duct is 50 mm. The ends of pretensioned tendons do not usually require any cover and are cut off flush with the end of the concrete member.

3.11 Vibration

The fact that thinner members are used in prestressed concrete construction than in comparable reinforced concrete construction leads to the natural frequency of such structures being near enough to the frequency of the applied loading to cause problems of resonance in some cases. Example of structures where vibrations should be considered include foundations for reciprocating machinery, bridge beams, especially those in footbridges, long-span floors and structures subjected to wind-excited oscillations, such as chimneys.

A method is given in Warner and Faulkes (1979) for finding the natural frequency of most types of prestressed concrete members.

References

Abeles, P. W. and Bardhan-Roy, B. K. (1981) *Prestressed Concrete Designer's Handbook*, Viewpoint, Slough.

Clark, L. A. (1982) *Concrete Bridge Design to BS5400*, Cement and Concrete Association, London.

Concrete Society (1983) *Partial Prestressing*, Technical Report No. 23, London.

Rowe, R. E. *et al.* (1987) *Handbook to British Standard BS8110: 1985 Structural Use of Concrete*. Viewpoint, London.

Warner, R. F. and Faulkes, K. A. (1979) *Prestressed Concrete*, Pitman, Sydney.

Chapter 4

LOSS OF PRESTRESS FORCE

4.1 Introduction

In Chapter 3 it was shown that one of the design conditions is that of maximum bending moment and minimum prestress force at any section. It is important, therefore, to obtain an estimate of the minimum prestress force throughout the structure.

There are several factors which cause the force in the prestressing tendons to fall from the initial force imparted by the jacking system. Some of these losses are immediate, affecting the prestress force as soon as it is transferred to the concrete member. Other losses occur gradually with time. These short and long-term losses are summarized in Table 4.1.

Friction losses only affect post-tensioned members, and vary along the length of a member. Thus the resulting prestress force anywhere in a post-tensioned member not only varies with time but with the position in the member.

Experience with the production of prestressed concrete members will allow good estimates of the loss of prestress force to be made, but in the absence of such information estimates may be based on the recommendations given in the following sections. This information is, of necessity, general and approximate, and any given structure should be examined carefully to determine whether these recommendations are valid. Many successful highway bridges have been constructed in the USA using lump-sum estimates of the loss of prestress force, rather than by determining the contribution of each of the effects listed in Table 4.1.

High accuracy is rarely justified in determining the loss of prestress force and an accuracy of $\pm 10\%$ is sufficient for most purposes. The ultimate strength of prestressed concrete members is very little affected by the initial prestress

Table 4.1 Prestress losses.

Short-term	Long-term
Elastic shortening	Concrete shrinkage
Anchorage draw-in	Concrete creep
Friction	Steel relaxation

force. Also, in service there is a low probability of the member being subjected to the full dead and imposed load, and there are partial factors of safety incorporated in the allowable concrete stresses. All these factors indicate that a prestressed concrete member is able to tolerate a small variation of prestress force.

4.2 Elastic shortening

Consider a pretensioned member with an eccentric prestress force P_i transferred to it as shown in Fig. 4.1. At the level of the prestressing tendons, the strain in the concrete must equal the change in the strain of the steel.
Thus

$$f_{co}/E_c = \Delta f_p/E_s;$$
$$\therefore \Delta f_p = m f_{co}, \tag{4.1}$$

where $m = E_s/E_c$, the modular ratio, f_{co} is the stress in the concrete at the level of the tendons, Δf_p is the reduction in stress in the tendons due to elastic shortening of the concrete to which they are bonded, and E_s and E_c are the moduli of elasticity of the steel and concrete respectively. The stress in the concrete is given by

$$f_{co} = \frac{P_e}{A_c} + \frac{(P_e e)e}{I}$$

$$= \frac{P_e}{A_c}\left(1 + \frac{e^2}{r^2}\right) \tag{4.2}$$

where P_e is the effective prestress force after elastic shortening, A_c and I are the cross-sectional area and second moment of area of the concrete section respectively, and r is the radius of gyration, given by $r^2 = I/A_c$.
Also,

$$P_e = A_{ps}\,(f_{pi} - \Delta f_p), \tag{4.3}$$

where f_{pi} is the initial stress in the tendons and A_{ps} is their cross-sectional area. Although, strictly speaking, the right-hand side of Equation 4.3 is the

Fig. 4.1

force in the tendon, for no applied axial force on the section this must equal the force in the concrete. Combining Equations 4.1, 4.2 and 4.3 gives

$$f_{co} = \frac{f_{pi}}{\left[m + \dfrac{A_c}{A_{ps}(1 + e^2/r^2)} \right]}. \tag{4.4}$$

If the tendons are closely grouped in the tensile zone, the loss due to elastic shortening may be found with sufficient accuracy by taking f_{co} as the stress in the concrete at the level of the centroid of the tendons. If the tendons are widely distributed throughout the section, then the above approximation is no longer valid. In this case the influences of the tendons, or groups of tendons, should be determined separately and then superimposed to give the total effective prestress force.

For a post-tensioned member the change in strain in the tendons just after transfer is still equal to the strain in the concrete at the same level, even though the ducts have not been grouted and there is no bond between the steel and concrete. The loss of stress in the tendon is therefore still given by Equation 4.1. In practice, the force in post-tensioned members at transfer is not constant as shown in Fig. 4.1. However, it is sufficiently accurate to base the elastic shortening loss on the initial prestress force, P_i, assumed constant along the member.

The value of f_{co} in Equation 4.4 should reflect the fact that, in general, a member deflects away from its formwork during tensioning and the stress at any section is modified by the self weight of the member. The additional tensile stress at the level of the tendon is equal to $M_i e/I$, so that the total value of f_{co} is given by

$$f_{co} = \frac{f_{pi}}{\left[m + \dfrac{A_c}{A_{ps}(1 + e^2/r^2)} \right]} - \frac{M_i e}{I}. \tag{4.5}$$

The value of f_{co} will vary along a member, since generally both e and M_i will vary. In this case an average value of f_{co} should be assumed.

For a post-tensioned member with a single tendon, or with several tendons tensioned simultaneously, there is no elastic shortening loss, since jacking would proceed until the desired prestress force is reached. In the more usual, and more economical, case where the tendons are tensioned sequentially, after the first tendon the tensioning of any subsequent tendon will reduce the force in those already anchored, with the exception of the last tendon, which will suffer no loss.

While it is possible to determine the resulting forces in a group of tendons for a given sequence of tensioning, the amount of work involved may be large. An acceptable approximation is to assume that the loss in each tendon is equal to the average loss in all the tendons. The loss for the first tendon is approximately equal to $m f_{co}$ (in practice it is always less but approaches this value as the number of tendons increases), and the loss for the last tendon is zero, so that the average loss is $1/2mf_{co}$.

In the case of pretensioned tendons, it is usually assumed that the total force is transferred to the member at one time and that the elastic shortening loss is mf_{co}.

EXAMPLE 4.1 ■ ■

Determine the loss of prestress force due to elastic shortening for the beam shown in Fig. 4.2. Assume that $f_{pi} = 1239\,\text{N/mm}^2$, $A_{ps} = 2850\,\text{mm}^2$, and $m = 7.5$.

Section properties:

$$w = 9.97\,\text{kN/m}; \; A = 4.23 \times 10^5\,\text{mm}^2; \; I = 9.36 \times 10^{10}\,\text{mm}^4; \; r = 471\,\text{mm};$$

At midspan:

$$M_i = 9.97 \times 20^2/8 = 498.5\,\text{kN m}$$

$$f_{co} = \frac{1239}{\left[7.5 + \dfrac{4.23 \times 10^5}{2850(1 + 558^2/471^2)} \right]} - \frac{498.5 \times 10^6 \times 558}{9.36 \times 10^{10}} = 14.97\,\text{N/mm}^2.$$

At the supports,

$$f_{co} = \frac{1239}{(7.5 + 4.23 \times 10^5/2850)} = 7.95\,\text{N/mm}^2.$$

Thus, in Equation 4.1,

$$\Delta f_p = \tfrac{1}{2} \times 7.5 \times (14.97 + 7.95)/2 = 43\,\text{N/mm}^2,$$

which represents a loss of 3.5% of initial stress.

■ ■

For pretensioned members, and for post-tensioned members once the ducts have been grouted, the prestress force is effectively held constant. Any bending moment at a section will induce extra stresses in the steel and concrete due to

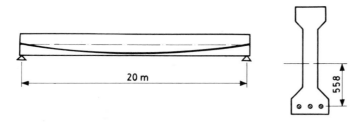

20 m

558

Fig. 4.2

composite action between the two materials, but the prestress force, as measured by the actual force transmitted to the ends of the member *via* the tendons, remains unaltered. For unbonded members, the prestress force will vary with the loading on the member, but in practice this effect is ignored.

4.3 Friction

In post-tensioned members there is friction between the prestressing tendons and the inside of the ducts during tensioning. The magnitude of this friction depends on the type of duct-former used and the type of tendon. There are two basic mechanisms which produce friction. One is the curvature of the tendons to achieve a desired profile, and the other is the inevitable, and unintentional, deviation between the centrelines of the tendons and the ducts.

A small, but finite, portion of a steel cable partly wrapped around a pulley is shown in Fig. 4.3(a). Since there is friction between the cable and the pulley, the forces in the cable at the two ends of the section are not equal. The frictional force is equal to μN, where μ is the coefficient of friction between cable and pulley. The triangle of forces for the short length of cable Δs is shown in Fig. 4.3(b); for the small angle $\Delta \alpha$, $N = T\Delta \alpha$. Thus, considering the horizontal equilibrium of the length of cable Δs,

$$T\cos (\Delta\alpha/2) + F = (T - \Delta T) \cos (\Delta\alpha/2).$$

For the small angle $\Delta\alpha$, $\cos (\Delta\alpha/2) \approx 1$.

$$\therefore \ T + F = T - \Delta T;$$
$$\therefore \ \mu T \Delta\alpha = - \Delta T.$$

Thus, in the limit as $\Delta s \rightarrow 0$,

$$\mathrm{d}T/\mathrm{d}\alpha = - \mu T.$$

The solution of this is

$$T(\alpha) = \mathrm{e}^{-\mu\alpha} \equiv \exp (-\mu\alpha)$$

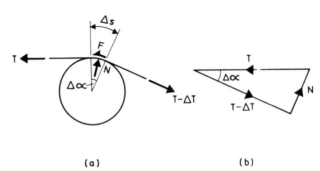

(a)   (b)

Fig. 4.3 Friction in a cable.

or

$$T_f = T_i \exp(-\mu\alpha_0),$$ (4.6)

where T_i and T_f represent the initial and final cable tensions respectively for a length of cable undergoing an angle change α_0.

The variation in tension in a tendon undergoing several changes of curvature, as shown in Fig. 4.4, may be described using Equation 4.6. For the first portion of the curve, with radius of curvature r_{ps1}, the force in the tendon at point 2 is given by

$$\begin{aligned}
P_2 &= P_1 \exp(-\mu\alpha_1) \\
&= P_1 \exp(-\mu s_1/r_{ps1}),
\end{aligned}$$

where s_1 is the length of the tendon to point 2. The force in the tendon has been denoted by P since it is the force in the concrete that is used in design. As noted previously, for no applied axial load the forces in the tendon and concrete must be equal. For most tendon profiles, s may be taken as the horizontal projection of the tendon, so that

$$P_2 = P_1 \exp(-\mu L_1/r_{ps1}).$$

For the portion of the tendon 2–3, the initial force is P_2, and the final force P_3 is given by

$$\begin{aligned}
P_3 &= P_2 \exp[-\mu(L_2/r_{ps2})] \\
&= P_1 \exp[-\mu(L_1/r_{ps1} + L_2/r_{ps2})].
\end{aligned}$$

This process can be repeated for all the changes in curvature along the length of the tendon. The tension in a curved tendon at an intermediate point along the

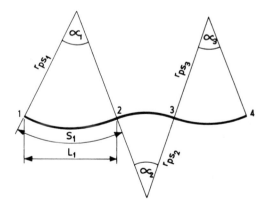

Fig. 4.4 Tendon with several curvature changes.

Fig. 4.5 The 'wobble' effect.

curved length is given by

$$P(x) = P_i \exp(-\mu x / r_{ps}), \tag{4.7}$$

where x is the distance from the start of the curve and P_i and $P(x)$ are the tendon forces at the beginning and end of the curve respectively.

Only variations of curvature in the vertical plane have so far been considered, but in many large bridge decks tendons curve in the horizontal plane as well, and the friction losses for these curvatures must also be taken into account.

The variation between the actual centrelines of the tendon and duct is known as the 'wobble' effect (Fig. 4.5). This is generally treated by considering it as additional angular friction, so that the expression for the force in a tendon due to both angular friction and wobble is given by

$$P(x) = P_i \exp[-(\mu x / r_{ps} + Kx)], \tag{4.8}$$

where K is a profile coefficient with units of $(\text{length})^{-1}$. The value of K depends on the type of duct used, the roughness of its inside surface, and how securely it is held in position during concreting.

The form of Equation 4.8 is equivalent to those given in BS8110, which also recommends a minimum value for K of 33×10^{-4} per metre, but if the ducts are securely held in position during concreting this may be reduced to 17×10^{-4} per metre. For greased strands wrapped in plastic sleeves, as used in slabs, K may be taken as 25×10^{-4} per metre. If $(\mu x / r_{ps} + Kx) \leqslant 0.2$, then Equation 4.8 may be simplified to

$$P(x) = P_i [1 - (\mu x / r_{ps} + Kx)]$$

Table 4.2 Coefficients of friction for different tendon types.

Surface-to-surface condition	Coefficient of friction
Lightly-rusted strand on unlined concrete duct.	0.55
Lightly-rusted strand on lightly-rusted steel duct.	0.30
Lightly-rusted strand on galvanized duct.	0.25
Bright strand on galvanized duct.	0.20
Greased strand on plastic sleeve.	0.12

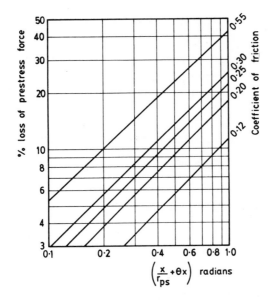

Fig. 4.6 Friction losses.

Typical values of μ for wires and strands against different surfaces are shown in Table 4.2. These values may be reduced to as low as 0.1 by lubricating the tendons prior to threading them into the ducts. In this case the coefficients must be substantiated by test results. The effect on the bond between steel and concrete if the ducts are subsequently grouted must also be considered.

An alternative to Equation 4.8 is to replace the term due to the wobble effect by an equivalent additional curvature θ radians per unit length, so that Equation 4.8 can now be written as

$$P(x) = P_i \exp\left[-\mu(x/r_{ps} + \theta x)\right]. \tag{4.9}$$

Typical values for θ are in the range 0.005–0.010 radians per metre. A graphical method of solving Equation 4.9 is shown in Fig. 4.6.

EXAMPLE 4.2 ■■

For the beam in Example 4.1 determine the prestress loss due to friction at the centre and the right-hand end if the prestress force is applied at the left-hand end. Assume $\mu = 0.25$ and $K = 17 \times 10^{-4}$ per metre.

The total angular deviation in a parabolic curve may be conveniently determined using the properties of the parabola shown in Fig. 4.7.

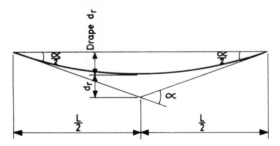

Fig. 4.7 Properties of parabolic profiles.

Thus, for the tendon profile in Fig. 4.2,

$$\alpha = 2 \tan^{-1}(4d_r/L)$$
$$= 2 \tan^{-1}(4 \times 558/20000)$$
$$= 0.222 \text{ radians.}$$

The radius of curvature is given by

$$r_{ps} = (d^2y/dx^2)^{-1} = L^2/8d_r$$
$$= 20^2/(8 \times 0.558)$$
$$= 89.61 \text{ m.}$$
$$P_i = 2850 \times 1239 \times 10^{-3}$$
$$= 3531.2 \text{ kN.}$$

Thus, using Equation 4.8,

$$P(x) = 3531.2 \exp[-(0.25x/89.61 + 17 \times 10^{-4}x)].$$

At midspan,

$$P(x = 10) = 3531.2 \exp[-(0.25 \times 10/89.61 + 17 \times 10^{-4} \times 10)]$$
$$= 3376.2 \text{ kN.}$$

Thus the loss is 155.0 kN, which is 4.4% of the initial force.
At the right-hand end,

$$P(x = 20) = 3531.2 \exp[-(0.25 \times 0.222 + 17 \times 10^{-4} \times 20)]$$
$$= 3228.9 \text{ kN.}$$

The loss is now 302.3 kN, that is, 8.6% of the initial force.

■ ■

The friction losses in the relatively shallow tendon in Example 4.2 are small, but in members with tendons of larger curvature the losses may be so large that the member must be tensioned from both ends to achieve an acceptable value of prestress force at the centre. In members with many tendons, it is the usual practice to tension half the number of tendons from one end and the remainder from the opposite end, resulting in the same net prestress force at midspan but a more even distribution of prestress force along the member than if all the tendons had been tensioned from the same end.

EXAMPLE 4.3   ■ ■

For the beam in Fig. 4.8, determine the minimum effective prestress force if an initial prestress force of 3000 kN is applied (i) at the left-hand end only; (ii) at both ends. Assume the same values of μ and K as in Example 4.2.

(i) The total angular change for the full length of the tendon is given by

$$\Sigma \frac{x}{r_{ps}} = \frac{18.75}{121.93} + \frac{12.5}{77.39} + \frac{18.75}{121.93}$$
$$= 0.469 \text{ radians.}$$

The minimum prestress force occurs at the right-hand end of the beam.

$$T(x = 50) = 3000 \exp[-(0.25 \times 0.469 + 17 \times 10^{-4} \times 50)]$$
$$= 2450.7 \text{ kN.}$$

Thus the loss is 549.3 kN, which is 18.3% of the initial force.

(ii) If the beam is tensioned from both ends, the minimum prestress force is at the centre of the beam. Then

$$\Sigma \frac{x}{r_{ps}} = \frac{18.75}{121.93} + \frac{6.25}{77.39}$$
$$= 0.235 \text{ radians.}$$
$$\therefore T(x = 25) = 3000 \exp[-(0.25 \times 0.235 + 17 \times 10^{-4} \times 25)]$$
$$= 2711.1 \text{ kN.}$$

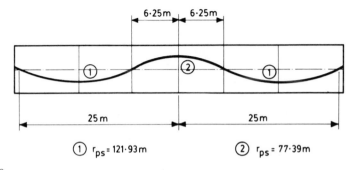

Fig. 4.8

The loss is now 288.9 kN, i.e. 9.6% of the initial force.

The frictional losses in the right-hand span have been greatly reduced by tensioning from both ends, although the prestress force at the centre support is the same in both cases.

■ ■

There are two additional frictional effects which occur. The first takes place as the tendons pass through the anchorages. This effect is small, however, of the order of 2%, and is usually covered by the calculated duct friction losses, which tend to be conservative. There is also a small amount of friction within the jack itself, between the piston and the jack casing, which will cause the load applied to the tendon to be smaller than indicated by the hydraulic pressure within the jack. This is usually determined by the jack manufacturer and compensation made in the pressure gauge readings.

Although friction is a cause of loss of prestress force principally in post-tensioned members, in pretensioned members there is some loss if the tendons are tensioned against deflectors, caused by friction between the tendon and the deflector. The magnitude of this loss will depend upon the details of the deflector, and will usually be determined from tests on the particular deflection system being used.

Further information on friction during tensioning may be found in a report of the Construction Industry Research and Information Association (1978).

4.4 Anchorage draw-in

A prestressing tendon may undergo a small contraction during the process of transferring the tensioning force from the jack to the anchorage; this is known as anchorage *draw-in*. The exact amount of this contraction depends on the type of anchorage used and is usually specified by the manufacturer of the anchorage. In the case of pretensioning it can be compensated easily by initially over-extending the tendons by the calculated amount of the anchorage draw-in.

Many anchorage systems use wedges to grip the tendon and transfer the tendon force to a solid steel anchorage set in the concrete. There is some deformation of the solid anchorage itself, but this is very small, and most of the contraction in length of the tendon takes place as a result of slip between the tendon and the wedges. A typical value would be 5 mm. The slip of the wedges can be reduced by ensuring that they are pushed forward as far as possible to grip the tendons before releasing the jack. Those anchorages which transfer the prestress force through a bar with a threaded nut or a wire with a shaped end suffer negligible draw-in.

Since the anchorage draw-in is a fixed amount which is dependent only on the type of anchorage used, the effect is much greater on a short prestressed concrete member than on a long one. However, the effect is greatly reduced in post-tensioned members by the friction that exists between the tendons and the ducts as the tendons move back due to the draw-in. The length of tendon used to determine the loss of prestress is not the total length of the tendon, but a smaller effective length, as shown in the next section.

4.5 Variation of initial prestress force along a member

It is now possible to look at how the prestress force varies along the length of a post-tensioned member immediately after transfer of the prestress force (ignoring elastic shortening). The line ABC in Fig. 4.9 represents the variation in prestress force away from the anchorage, based on Equation 4.8.

The vertical ordinate AD represents the loss of prestress force due to draw-in, ΔP_A. Over the length AB, the tendons are being relaxed, so that they tend to move in the opposite direction to the original direction moved during tensioning. On the assumption that Equation 4.8 applies regardless of which way the tendons are actually moving, the variation of prestress force will follow the curve DB, which is the reflection of the curve AB. Beyond point B the force in the tendon is unaffected by the draw-in. For most shapes of tendon profile used, the total deviated angle is small, and the two curves AB and DB in Fig. 4.9 may be approximated by straight lines, as shown in Fig. 4.10.

If the friction loss per metre is $p\,\text{kN/m}$, then from Fig. 4.10 it can be seen that

$$\Delta P_A/2 = px_A. \qquad (4.10)$$

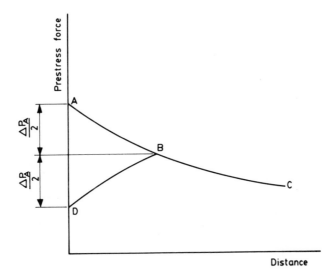

Fig. 4.9 Anchorage draw-in.

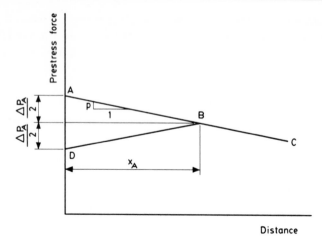

Fig. 4.10 Idealized prestress force distribution.

If the anchorage draw-in is δ_{ad}, then the reduction in stress is based on an effective length of the tendons, x_A, since beyond this length the tendon is unaffected by the draw-in.

Thus, the loss of stress in the steel is given by

$$\Delta f_p = \varepsilon_s E_s$$
$$= (\delta_{ad}/x_A)E_s.$$

Since $\Delta f_p A_{ps} = \Delta P_A/2$, then

$$\Delta P_A/2 = (\delta_{ad}/x_A)E_s A_{ps}.$$

Thus, in Equation 4.10,

$$p x_A = (\delta_{ad}/x_A)E_s A_{ps};$$
$$\therefore x_A = (\delta_{ad} E_s A_{ps}/p)^{1/2}. \tag{4.11}$$

EXAMPLE 4.4 ■■

For the beam in Example 4.1 determine the initial prestress force distribution along the beam if the anchorage draw-in is 5 mm. Assume $E_s = 195 \, \text{kN/mm}^2$.

The friction loss per unit length near the anchorages is given by

$$P = P_i\{1 - \exp[-(\mu/r_{ps} + K)]\}$$
$$= 3531.2\{1 - \exp[-(0.25/89.61 + 17 \times 10^{-4})]\}$$
$$= 15.82 \, \text{kN/m}.$$

Thus, in Equation 4.11,

$$x_A = (5 \times 195 \times 10^3 \times 2850/15.82)^{1/2} \times 10^{-3}$$
$$= 13.25\,\text{m}.$$

The loss of prestress froce at the left-hand end is given by

$$\Delta P_A = 2 \times 15.82 \times 13.25$$
$$= 419.3\,\text{kN}.$$

The total force variation in the tendon is now as shown in Fig. 4.11.

The prestress force at midspan is 3270.1 kN, representing a loss of 7.4%. In this example there would be no benefit from tensioning from the right-hand end as well, since in this case the prestress force distribution (shown by broken lines in Fig. 4.11) would result in a prestress force at the right-hand end which is less than that obtained if tensioning is carried out from the left-hand end only.

■ ■

4.6 Concrete shrinkage

Shrinkage of concrete was discussed in Chapter 2, and one of its effects in prestressed concrete members is that, since the prestressing steel is connected by bond or anchorage to the concrete, the steel also contracts, and the prestress force is reduced.

Shrinkage is dependent on many factors, and approximate long-term values for shrinkage strain to be used in design are given in BS8110. These are 100×10^{-6} for outdoor exposure in the UK and 300×10^{-6} for indoor exposure. For other conditions, shrinkage strains may be found from Fig. 2.6.

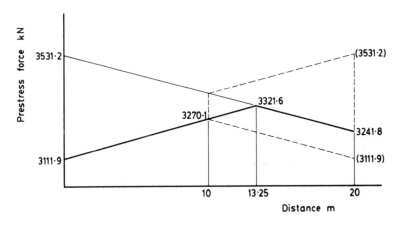

Fig. 4.11 Prestress force distribution for beam in Example 4.1.

4.7 Concrete creep

The phenomenon of creep in concrete was also discussed in Chapter 2, and its principal effect in prestressed concrete members is the same as that due to shrinkage, namely a reduction in prestress force caused by shortening of the member with time. As with shrinkage, there are many factors which affect the creep of concrete, and approximate values of creep strain to be used in estimating prestress losses due to creep are given in BS8110.

The specific creep strain is given by the formula

$$\text{specific creep} = \phi/E_{ci}, \tag{4.12}$$

where ϕ is a creep coefficient varying between 1.8 for transfer at 3 days and 1.4 for transfer at 28 days, for indoor and outdoor exposure in the UK, and E_{ci} is the modulus of elasticity of the concrete at transfer. In order to determine the creep strain from Equation 4.12, the stress in the concrete should be that adjacent to the tendons immediately after transfer.

Creep and shrinkage losses are very little affected by the grade of steel used. It is thus advantageous to use as high a grade of steel as possible, since the percentage losses of prestress force due to creep and shrinkage will be minimized.

For lightweight aggregate concretes, the creep and shrinkage effects are greater than, and the modulus of elasticity less than, those of normal-density concretes. Thus the loss of prestress force to be expected will be greater.

4.8 Steel relaxation

Steel relaxation is described in Chapter 2, along with the manufacturing processes used to minimize relaxation. The long-term relaxation loss is specified in BS8110 as the 1000-hour relaxation test value given by the tendon manufacturer or, in the absence of this, the value given in Table 2.2, multiplied by the factors given in Table 4.3. These factors include allowances for the effects of creep and shrinkage and in the case of pretensioned members, the effects of elastic shortening. In cases where the tendons are subjected to high temperature, due allowances for the increase in relaxation should be made.

EXAMPLE 4.5   ■ ■

For the beam in Example 4.1, determine the total prestress losses due to the following causes: (i) shrinkage; (ii) creep; (iii) steel relaxation.

(i) Assuming indoor conditions of exposure, from BS8110,

$$\varepsilon_{sh} = 300 \times 10^{-6}.$$

Table 4.3 Relaxation factors.

| | Wire and strand | | |
| | Relaxation class as defined in BS5896:1980 | | |
	1	2	Bar
Pretensioning	1.5	1.2	—
Post-tensioning	2.0	1.5	2.0

The loss of prestress is thus given by

$$\Delta f_p = \varepsilon_{sh} E_s$$
$$= 300 \times 10^{-6} \times 195 \times 10^3$$
$$= 59 \, \text{N/mm}^2.$$

(ii) From Table 2.1, with $f_{ci} = 0.67 f_{cu} = 26.8 \, \text{N/mm}^2$, $E_{ci} = 25 \, \text{kN/mm}^2$, and assuming that transfer takes place at 28 days,

$$\phi = 1.4/25 \times 10^3 = 56 \times 10^{-6}.$$

It can be shown that the stress in the concrete at the level of the tendons, f_{co}, based on an average f_{pi} of $1126 \, \text{N/mm}^2$, allowing for friction losses, is $13.28 \, \text{N/mm}^2$ at midspan and $7.22 \, \text{N/mm}^2$ at the support.

Thus the average value of $f_{co} = 1/2(13.28 + 7.22)$
$$= 10.25 \, \text{N/mm}^2.$$

Thus,

$$\Delta f_p = 56 \times 10^{-6} \times 195 \times 10^3 \times 10.25$$
$$= 112 \, \text{N/mm}^2.$$

(iii) Assuming that the prestressing tendons are of low-relaxation steel (BS5896, Class 2), with 1000-h relaxation of 2.5% at $0.7 f_{pu}$ (see Table 2.2), the long-term relaxation is given by

$$\Delta f_p = 1.5 \times 0.025 \times 1239$$
$$= 46 \, \text{N/mm}^2.$$

The total long-term losses are thus $217 \, \text{N/mm}^2$, or 17.5% of the initial prestress.

■ ■

4.9 Total prestress losses

If the initial prestress force applied to a member is P_i, then the effective prestress force at transfer is αP_i, while that at service load is βP_i. The value of α reflects the short-term losses due to elastic shortening, anchorage draw-in and friction, while β accounts for the long-term losses due to concrete creep and shrinkage and steel relaxation.

Although there are many factors which affect the total loss of prestress force, as described in the preceding sections, it is very useful at the initial design stage to have an approximate figure for the prestress loss. This can be refined later in the design process, when more details of the prestressing steel are available.

For both pretensioned and post-tensioned members, the values of α and β may be taken as 0.90 and 0.75 respectively. For the beam in Example 4.1, the corresponding values are 0.89 and 0.71.

Lump-sum estimates for losses are given in the American specification for bridges, AASHTO (1975). For losses other than those due to friction and anchorage draw-in, these are $310 \, \text{N/mm}^2$ and $228 \, \text{N/mm}^2$ for pretensioned and post-tensioned members respectively.

4.10 Measurement of prestress force

The actual force transmitted to the prestressing steel by the jack in a post-tensioned member is measured by a combination of measurement of the hydraulic pressure in the jack and measurement of tendon extension during tensioning. Most jacks are compensated for the small amount of friction between the piston and the jack casing. They are calibrated by the manufacturers against load cells and are usually accurate to within a few per cent.

Knowledge of the expected extension of the steel during tensioning serves as a check on the calculations for the loss of prestress force due to friction and elastic shortening. If the measured extension for a given jack hydraulic pressure deviates by more than 5% from the expected value, then corrective action should be taken.

If the measured extension is too low, then the friction effects have been underestimated and the prestress force along the member will be less than expected. Conversely, if the measured extension is too high, the friction effects have been overestimated and the actual prestress force along the member will be greater than expected. The first situation is potentially the more serious, and may be remedied by making a revised estimate of $(\mu/r_{ps} + K)$ based on the actual extension achieved, recalculating the extension required to give the desired prestress force along the member and continuing the tensioning until this extension is reached. The initial tendon force must not exceed $0.8 f_{pu} A_{ps}$, however, as described in the next section.

The measured extension may also indicate whether a blockage has occurred in the duct during tensioning. Thus, if only half the expected extension is achieved, even though the jack pressure indicates that the full force has been applied to the tendon, this could indicate that the tendon is not being tensioned uniformly along its length and that grout may have entered the duct during concreting, effectively holding a portion of the tendon rigid. This illustrates why prestress force measurement should not be based on tendon extension alone, since, in the example mentioned above, tensioning would have continued, possibly causing fracture of the tendon.

If a steel tendon with length L is tensioned gradually from zero tension up to the maximum force of $0.7f_{pu}A_{ps}$, the expected elongation, ignoring friction, is given by

$$\delta_e = 0.7f_{pu}L/E_s.$$

There is always some initial slack in a tendon, and the usual procedure is to apply a small force P', of the order of 10% of the final prestress force, and to measure the total extension from the initial extension due to this force. The extra elongation expected, δ_{ex}, when the tendon is tensioned to its full force, is then given by

$$\delta_{ex} = [(P_i - P')/P_i]\delta_e.$$

The elastic shortening of the concrete must be added to, and any anchorage draw-in at the untensioned, or dead-end anchorage deducted from, the calculated elongation.

In order to consider the effect of friction on the elastic elongation of the prestressing steel, consider a length L of steel tendon, subjected to a prestress force P_i, as shown in Fig. 4.12. The force in the tendon at a distance x from the tensioned end is given by Equation 4.8, that is

$$P(x) = P_i \exp[-(\mu x/r_{ps} + Kx)].$$

For a small length of tendon Δx, the extension is given by

$$\Delta\delta = (P(x)/E_s A_{ps})\Delta x.$$

Thus, for the total length L, the extension is given by

$$\delta = \int_0^L \frac{P(x)}{E_s A_{ps}}dx$$

$$= (P_i/E_s A_{ps})\int_0^L \exp[-(\mu/r_{ps} + K)x]dx$$

$$= \frac{-P_i\{\exp[-(\mu/r_{ps} + K)L] - 1\}}{E_s A_{ps}(\mu/r_{ps} + K)}. \tag{4.13}$$

EXAMPLE 4.7 ■■

Determine the measured elongation for the beam in Example 4.1.

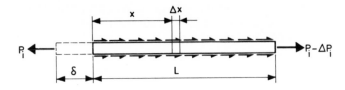

Fig. 4.12 Tendon elongation.

From Equation 4.13,

$$\delta_e = \frac{-3531.2\{\exp[-(0.25/89.61 + 17 \times 10^{-4}) \times 20] - 1\}}{195 \times 10^6 \times 2850 \times 10^{-6}(0.25/89.61 + 17 \times 10^{-4})}$$

$$= 0.122\,\text{m, or } 122\,\text{mm.}$$

Ignoring the frictional resistance, $\delta_e = 127\,\text{mm}$.

■■

4.11 Initial overtensioning

One way of overcoming the losses of prestress force due to anchorage draw-in and friction is to tension the tendons initially to a stress level greater than the usual $0.7f_{pu}$. It is stipulated in BS8110 that, for mature concrete the maximum initial stress is $0.8f_{pu}$, and the usual procedure is to check that the stress in the tendons is nowhere greater than $0.7f_{pu}$ once the immediate losses have occurred. This stress may be increased to $0.75f_{pu}$ if necessary.

EXAMPLE 4.8   ■■

Determine the initial prestress force distribution for the beam in Example 4.1 if the tendons are initially tensioned to $0.8f_{pu}$.

$$\text{Initial prestress force} = 0.8 \times 1770 \times 2850 \times 10^{-3}$$

$$= 4035.6\,\text{kN.}$$

Thus the friction loss per metre, p, is given by

$$p = 4035.6\{1 - \exp[-(0.25/89.61 + 17 \times 10^{-4})]\}$$

$$= 18.08\,\text{kN/m.}$$

Thus, in Equation 4.11,

$$x_A = (5 \times 195 \times 10^3 \times 2850/18.08)^{1/2} \times 10^{-3}$$

$$= 12.40\,\text{m.}$$

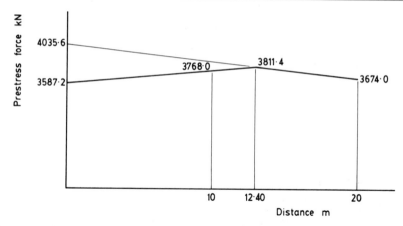

Fig. 4.13 Initial over-tensioning for beam in Example 4.1.

The loss of prestress force at the anchorage is thus given by

$$\Delta P_A = 2 \times 18.08 \times 12.40$$
$$= 448.4 \, \text{kN}.$$

The prestress force distribution is as shown in Fig. 4.13. The effective prestress force at the tensioning end is 3587.2 kN, corresponding to $0.71 f_{pu}$, and the prestress force at midspan is 3768.0 kN, which is a 15.2% increase on the midspan value obtained if the tendons are initially tensioned to $0.7 f_{pu}$.

■ ■

Further information on prestress losses in general may be found in Abeles and Bardhan-Roy (1981) and Rowe *et al.* (1987), and on the behaviour of tendons during tensioning in Fédération Internationale de la Précontrainte (1986).

References

Abeles, P. W. and Bardhan-Roy, B. K. (1981) *Prestressed Concrete Designer's Handbook*, Viewpoint, Slough.

American Association of Highway and Transportation Officials (1975) *AASHTO Interim Specifications – Bridges*, Subcommittee on Bridges and Structures, Washington.

Construction Industry Research and Information Association (1978) *Prestressed Concrete – Friction Losses During Stressing*. Report No. 74 CIRIA, London.

Fédération Internationale de la Precontrainte (1986): *Tensioning of Tendons: Force Elongation Relationship*.

Rowe, R. E. *et al.* (1987) *Handbook to British Standard BS8110:1985 Structural use of Concrete*, Viewpoint, London.

Chapter 5

ANALYSIS OF SECTIONS

5.1 Introduction

The design of a prestressed concrete structure involves many considerations, the most important of which is the determination of the stress distributions in the individual members of the structure. In most types of structure it is usually sufficient to consider certain critical sections where the stresses are greatest. In prestressed concrete structures, however, since high stresses are introduced by the prestress force, all sections must be considered as critical and the stress distributions checked for all stages of loading. The practical means of carrying this out will be discussed in Chapter 9.

This chapter is concerned with the distribution of flexural stresses at the serviceability and ultimate limit states. These two distributions are different, but in determining them the three basic principles employed are the same. They are:

(a) Strain distribution

This is assumed to be linear in elastic bending theory, and this assumption is also found to be sufficiently true for concrete members even up to the point of failure. The strain in the steel in pretensioned and bonded post-tensioned members is assumed to be the same as that in the concrete at the same level.

(b) Material stress–strain curves

These have been described in detail in Chapter 2 for both steel and concrete.

(c) Equilibrium

At any section in a prestressed concrete member there must be equilibrium between the stress resultants in the steel and concrete and the applied bending moment and axial load (if any) at that section.

The basic difference between the analysis of sections at the serviceability and ultimate limit states is that in principle (b) different regions of the stress–strain curves are used in each case.

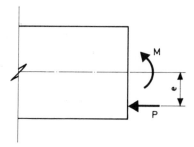

Fig. 5.1 Prestress moment at a section.

5.2 Serviceability limit state

The analysis of sections in Class 1 and 2 members at the serviceability limit state is carried out by treating the section as linearly elastic and using ordinary bending theory. (The analysis for Class 3 members will be considered in Section 5.11.) This is justified by the fact that, at the service load, the stress–strain curve for steel is linear, and that for concrete is approximately so. Furthermore, Class 1 and 2 prestressed concrete members remain uncracked at service loads, justifying the use of a value of second moment of area based on the gross concrete section.

In Section 1.3 it was shown that, in an unloaded prestressed concrete member, at any cross-section the concrete behaves as if it were subjected to an axial force P and a bending moment Pe, where e is the eccentricity of the prestress force at that section (Fig. 5.1). Thus the stress distribution due to an eccentric prestress force can be written as:

$$f_t = (P/A_c) - (Pe/Z_t),$$
$$f_b = (P/A_c) + (Pe/Z_b),$$

where e is taken as positive if it is below the member centroidal axis, A_c is the cross-sectional area and where f_t and f_b are the stresses and Z_t and Z_b the section moduli for the top and bottom fibres of the member respectively. The sign convention used is that compressive stresses are positive, as are sagging bending moments. If now an external sagging bending moment M, is applied to the section, an additional distribution of stresses is introduced and the resultant stress distribution due to prestress force and applied bending moment may be found by superposition.

$$f_t = (P/A_c) - (Pe/Z_t) + (M/Z_t) \tag{5.1a}$$
$$f_b = (P/A_c) + (Pe/Z_b) - (M/Z_b). \tag{5.1b}$$

If in addition to the applied bending moment at the section there is also an applied axial load, then the force P in Equations 5.1(a) and (b) is the sum of the prestress force and the applied axial load.

EXAMPLE 5.1 ■ ■

A simply supported pretensioned concrete beam has dimensions as shown in Fig. 5.2 and spans 15 m. It has an initial prestress force of 1100 kN applied to it and it carries a uniformly distributed imposed load of 12 kN/m. Determine the extreme fibre stresses at midspan (i) under the self weight of the beam, if the short-term losses are 10% and the eccentricity is 325 mm below the beam centroid; (ii) under the service load, when the prestress force has been reduced by a further 10%.

For the beam cross-section:

$$A_c = 2.13 \times 10^5 \text{ mm}^2$$
$$Z_b = Z_t = 35.12 \times 10^6 \text{ mm}^3$$
$$w = 5.1 \text{ kN/m}$$
$$M_i = 5.1 \times 15^2/8 = 143.4 \text{ kN m}$$
$$M_s = 17.1 \times 15^2/8 = 480.9 \text{ kN m}$$
$$\alpha P_i = 0.9 \times 1100 = 990 \text{ kN}$$
$$\beta P_i = 0.8 \times 1100 = 880 \text{ kN.}$$

(i)
$$f_t = \frac{990 \times 10^3}{2.13 \times 10^5} - \frac{990 \times 10^3 \times 325}{35.12 \times 10^6} + \frac{143.4 \times 10^6}{35.12 \times 10^6}$$
$$= 4.65 - 9.16 + 4.08$$
$$= -0.43 \text{ N/mm}^2.$$
$$f_b = 4.65 + 9.16 - 4.08$$
$$= 9.73 \text{ N/mm}^2.$$

(ii)
$$f_t = \frac{880 \times 10^3}{2.13 \times 10^5} - \frac{880 \times 10^3 \times 325}{35.12 \times 10^6} + \frac{480.9 \times 10^6}{35.12 \times 10^6}$$
$$= 4.13 - 8.14 + 13.69$$
$$= 9.68 \text{ N/mm}^2.$$
$$f_b = 4.13 + 8.14 - 13.69$$
$$= -1.42 \text{ N/mm}^2.$$

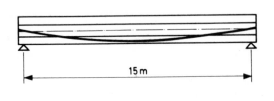
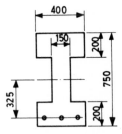

Fig. 5.2

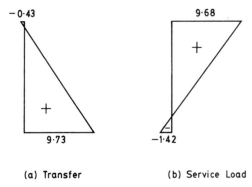

(a) Transfer   (b) Service Load

Fig. 5.3 Stress distribution for beam in Example 5.1 (N/mm^2).

The two stress distributions, at transfer and service load, are shown in Fig. 5.3.
■ ■

The stress distributions shown in Fig. 5.3 are typical of those in a prestressed concrete member under maximum and minimum loads, and illustrate the point made in Chapter 1 that an important difference from reinforced concrete is that, with prestressed concrete the minimum load condition is always an important one. These four stress conditions lead to a method of design for prestressed concrete sections which will be discussed in further detail in Chapter 9.

So far the prestress force in a prestressed concrete member has been considered to be provided by one layer of tendons, so that the resultant prestress force coincides with the physical location of the layer of tendons at each section. However, there is usually more than one layer of tendons in prestressed concrete members. In this case the resultant prestress force coincides with the location of the resultant of all the individual prestressing tendons, even if it is not physically possible to locate a tendon at this position.

For post-tensioned members where the duct diameter is not negligible in comparison with the section dimensions, due allowance for the duct must be made when determining the member section properties. For pretensioned members the transformed cross-section should, strictly speaking, be used. However, in practice, the section properties are generally determined on the basis of the gross cross-section.

5.3 Additional steel stress due to bending

In the case of ungrouted post-tensioned members there is no bond between the prestressing steel and the surrounding concrete, but with pretensioned and grouted post-tensioned members bond is present, and bending of the member induces stress in the steel, as in a reinforced concrete member. It is the bond between the steel and concrete which makes the ultimate load behaviour of pretensioned and grouted post-tensioned members very similar to that of reinforced concrete members, and different from that of ungrouted post-tensioned members. The bond enables composite behaviour between the steel and concrete to take place, and the extra stresses induced in the steel at the serviceability limit state may be determined by using the transformed cross-section properties.

EXAMPLE 5.2 ■■

The beam in Example 5.1 is pretensioned with tendons having a total cross-sectional area of 845 mm^2. Determine the stress in the tendons under the service load.

All of the loads acting on the beam are resisted by the *transformed* concrete section, as shown in Fig. 5.4. The transformed area of the prestressing steel is mA_{ps}, where m is the modular ratio E_s/E_c. For typical values of E_s and E_c of 195 kN/mm^2 and 28 kN/mm^2 respectively, m is approximately 7.0.

It can be shown that $\bar{y} = 384$ mm and that the second moment of area of the transformed section is 1.38×10^{10} mm^4. The eccentricity about the centroid of the transformed section is 316 mm.

The steel stress induced by the service load is given by

$$\Delta f_{ps} = m(M_s - M_i)y/I$$
$$= 7.0 \times (480.9 - 143.4) \times 10^6 \times 316/(1.38 \times 10^{10})$$
$$= 54 \text{ N/mm}^2.$$

The effective steel prestress after all losses have occurred is given by

$$f_{pe} = 880 \times 10^3/845$$
$$= 1041 \text{ N/mm}^2$$

and the total steel stress f_{pb} is now $1041 + 54 = 1095$ N/mm^2. The extra stress induced by bending in this, and most cases, is thus small, and is usually ignored. ■■

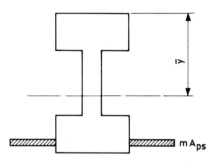

Fig. 5.4 Transformed section.

5.4 Post-cracking behaviour

If the service load on the beam is increased, then the tensile stress at the soffit of the beam will increase proportionately, until the modulus of rupture is reached. If this is, say, 3.6 N/mm² for the concrete in the beam in Example 5.1, then the bending moment M_{cr} which will cause this stress to be reached is given by

$$-3.6 = \frac{880 \times 10^3}{2.13 \times 10^5} + \frac{880 \times 10^3 \times 325}{35.12 \times 10^6} - \frac{M_{cr} \times 10^6}{35.12 \times 10^6};$$

$$\therefore M_{cr} = 557.5 \, \text{kN m}.$$

If the service load bending moment is increased beyond this value, the concrete in the tensile zone must be assumed to have cracked and the stresses in the section must be found using a *cracked* transformed section, the contribution of all concrete below the neutral axis being neglected. It is assumed that there is still bond between the steel and concrete, even though the concrete surrounding the steel is cracked. This is the same assumption that is made in reinforced concrete section analysis.

The strain and stress diagrams for a cracked section are shown in Fig. 5.5, and the procedure for a cracked-section analysis is as follows:

(a) Choose a strain in the concrete extreme fibres, ε_c;
(b) Choose a neutral axis depth, x;
(c) Determine the concrete and steel stresses from the relevant stress–strain curves, neglecting the concrete in tension below the neutral axis;
(d) Check whether total compression equals total tension within the section (for no applied axial load). If it does, determine the moment of resistance of the section. If not, go back to steps (a) and (b) and repeat steps (c) and (d);
(e) Repeat steps (a)–(d) until the moment of resistance equals the applied bending moment.

Note that when determining the force in the prestressing steel, the total strain ε_{ps} comprises two components, that due to the bending action of the beam ε_p, and that due to the effective prestrain in the tendon, ε_{pe}.

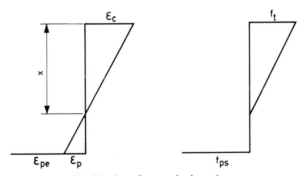

Fig. 5.5 Strain and stress distributions for cracked section.

A cracked-section analysis is extremely laborious, since there are two unknowns, ε_c and x, and is best carried out by programmable calculator or microcomputer. For a given value of ε_c, x may be found from consideration of internal equilibrium. The resulting moment of resistance is then compared with the applied bending moment. The procedure is exactly the same as that which will be used in the next section to analyse a member at the ultimate load. There, the value of ε_c is fixed, leaving the only unknown as x, which simplifies the calculations considerably.

Cracked-section analyses are useful in determining the steel stresses in Class 3 members, and these are used as a means of checking that the crack widths are not excessive (see Section 5.11).

EXAMPLE 5.3   ■■

For the beam in Example 5.1 use a cracked-section analysis for an applied bending moment of 557.5 kN m to determine the stress in the steel.

$$\varepsilon_{pe} = \frac{880 \times 10^3}{845 \times 195 \times 10^3} = 0.00534.$$

The stress and strain distributions which give equal tension and compression and also balance the applied bending moment are found using the trial and error approach outlined above, and are shown in Fig. 5.6.

From Fig. 5.6, the moment of resistance is given by

$$M_r = 931.4 \times 0.599$$
$$= 557.9 \text{ kN m};$$
$$f_{ps} = (0.000312 + 0.00534) \times 195 \times 10^3$$
$$= 1102 \text{ N/mm}^2.$$

■■

It has been assumed in the foregoing that the concrete stress–strain curve is

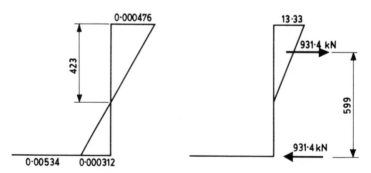

Fig 5.6 Strain and stress distributions for Example 5.3.

linear. Although this is a reasonable approximation for the initial region of the stress–strain curve, for the analysis of a cracked section at much higher loads the full non-linear curves for both steel and concrete must be considered.

5.5 Ultimate load behaviour

As the service load is increased still further, both the concrete and steel stresses will increase, following the respective stress–strain curves. These are shown in Fig. 5.7, and show the actual values that occur in the materials. These curves differ from those shown in Figs 3.3 and 3.4, where the *design* stress–strain curves are given. At present, the *actual* behaviour of the beam is being considered, with γ_m for both concrete and steel taken as 1.0. The design, or allowable, ultimate load on the beam, which incorporates partial factors of safety, will be considered later.

As the applied load is increased, the strain and stress distributions across the section change, the strain distribution remaining linear as described in Section 5.1. The stress in the extreme fibres of the beam section follows the stress–strain curve in Fig. 5.7(a), and will eventually reach the limiting value of $0.67f_{cu}$, although the extreme-fibre concrete strain continues to increase, until it reaches its maximum value, ε_{cu}, of 0.0035. This strain has been found to be the average maximum that concrete of all grades can withstand before crushing of the material. At all times the total compression in the concrete and the tension in the steel are equal (for no applied axial load) and the moment of resistance is always given by Cz or Tz (Fig. 5.8).

By the time the limiting concrete strain has been reached, the total strain in the prestressing steel ε_{pb} can either be (a) *greater* than ε_2, in which case the steel will have yielded before the concrete finally crushes – a *ductile* failure (such a beam is termed *under-reinforced*); or (b) *less* than ε_2, in which case the steel will

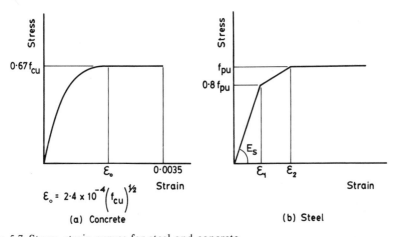

Fig. 5.7 Stress–strain curves for steel and concrete.

not have yielded before the concrete finally crushes – a *brittle* failure (such a beam is termed *over-reinforced*). If the steel strain equals ε_2, then the section is said to be *balanced*.

This situation is analogous to that in reinforced concrete members at the ultimate limit state, the only difference being that the initial strain ε_{pe} in the steel must be considered. As with reinforced concrete members it is the ductile failure which is desirable, since it is gradual and gives ample warning. The load–deflection curves for typical under- and over-reinforced members are shown in Fig. 6.10.

EXAMPLE 5.4 ■ ■

Determine the ultimate applied load that the beam in Example 5.1 can support if $f_{cu} = 40 \text{ N/mm}^2$ and $f_{pu} = 1860 \text{ N/mm}^2$.

The procedure is similar to that used in the cracked-section analysis of Example 5.3, except that now the extreme-fibre concrete strain is fixed at $\varepsilon_{cu} = 0.0035$. The stress and strain distribution determined from Figs. 5.7(a) and (b) are shown in Fig. 5.8.

$$\varepsilon_{pb} = 0.00534 + [(d - x)/x]0.0035,$$

where d is the effective depth of the tendon.

With the value of ε_c fixed at $\varepsilon_{cu} = 0.0035$, x is determined by considering internal equilibrium. With irregular sections it may be easier to determine x by trial and error rather than to find it directly. Note that the concrete stress block is made up of two portions, corresponding to the two portions of the concrete stress–strain curve, one rectangular and one parabolic. The dimensions of the stress block are dependent on the characteristic strength of the concrete, and are shown in Fig. 5.8 for $f_{cu} = 40 \text{ N/mm}^2$. The derivation of the dimensions of the stress block may be found in Mosley and Bungey (1987).

The resulting strain and stress distributions are shown in Fig. 5.9. The

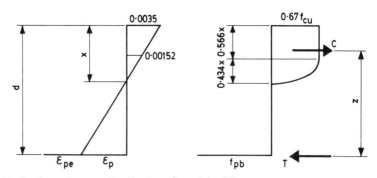

Fig. 5.8 Strain and stress distributions from Fig. 5.7.

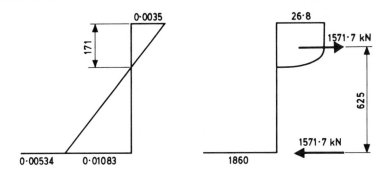

Fig. 5.9 Strain and stress distributions for Example 5.4.

ultimate moment of resistance of the section is then given by

$$M_u = 1571.7 \times 0.625$$
$$= 983.3 \, \text{kN m}.$$

This corresponds to a uniformly distributed load of 34.9 kN/m. The steel strain for this neutral axis depth is greater than the yield strain, so this is an example of a ductile, under-reinforced section.

■ ■

It should be noted here that the load of 34.9 kN/m is that which would cause *actual* failure of the beam. What is usually required is the maximum safe, or *design*, load that can be supported. This will be considered in Section 5.7. The full analysis of the behaviour of a prestressed concrete beam from transfer of prestress force to ultimate load has been shown in Examples 5.1–5.4 to illustrate the basic approaches to the analysis. For uncracked sections, an analysis by elastic bending theory can be used. Once the concrete has cracked, this theory is still applicable, except that the cracked-section properties must be considered. Once the concrete stress in the extreme fibres approaches the non-linear portion of the concrete stress–strain curve, then elastic bending theory is no longer valid, and the analysis must be carried out using the three basic principles stated in Section 5.1. Throughout, however, the equilibrium conditions between internal stress resultants and applied loads have been shown in order to emphasize the same basic behaviour of the beam at all levels of load.

5.6 Variation of steel stress

In order to obtain an overall view of the behaviour of the beam in Examples 5.1–5.4, it is useful to consider the variation of stress in the prestressing steel at the centre of the beam as the load on the beam is increased. This is shown in Fig. 5.10.

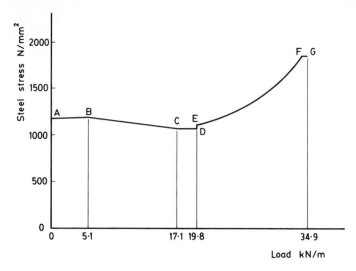

Fig. 5.10 Variation of prestress force for beam in Examples 5.1–5.4.

Assuming that the tendons are initially stressed to $0.70 f_{pu}$, that is $1302 \, \text{N/mm}^2$, and that the immediate losses due to elastic shortening and anchorage draw-in total 10%, then the initial stress in the steel at midspan is $1172 \, \text{N/mm}^2$, point A.

On removal of the falsework, or lifting of the beam from its casting bed, there is a slight increase in the steel stress as the beam begins to support its own weight, point B.

When the beam is in position and subjected to its service load of 17.1 kN/m, the prestress force is assumed to have been reduced by 10% due to long-term losses, but the bending moment at the section has increased, leading to a net steel stress of $1095 \, \text{N/mm}^2$, point C, as determined in Example 5.2.

As the load is increased beyond the service load, the stress increases slightly until point D, at which point the concrete in the extreme bottom fibres of the beam cracks, causing a sudden increase in steel stress to $1102 \, \text{N/mm}^2$, point E, as shown in Example 5.3. From this point onwards, the stress increases more rapidly as the neutral axis rises and the extreme-fibre concrete stress increases. Finally, the steel stress reaches the yield value of $1860 \, \text{N/mm}^2$, point F, after which it remains constant until point G, when failure occurs by crushing of the concrete.

A useful way of looking at the various stages in the behaviour of a prestressed concrete member is to look at its load–deflection curve. A typical such curve for a Class 1 member is shown in Fig. 5.11, along with the stress distributions in the member at each stage.

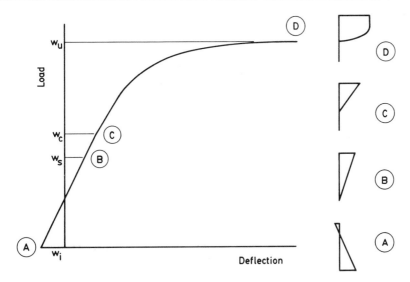

Fig. 5.11 A load–deflection curve for a Class 1 member.

5.7 Design ultimate strength

The uniform load of 34.9 kN/m determined in Example 5.4 is that which would cause physical collapse of the beam. What is more usually required is the *safe* ultimate load that the beam can support, that is the load which gives an adequate factor of safety against failure of the materials. This is found by introducing the partial factors of safety for the steel and concrete material properties described in Chapter 3. The stress–strain curves shown in Fig. 5.7 are now modified to those shown in Fig. 5.12 with γ_m for concrete and steel taken as 1.5 and 1.15 respectively.

The stress–strain curve for concrete leads to the strain and stress distributions within the section shown in Fig. 5.13. Again, the dimensions of the stress block are dependent on the characteristic strength of the concrete, and are shown in Fig. 5.13 for $f_{cu} = 40$ N/mm².

EXAMPLE 5.5   ■ ■

Determine the ultimate moment of resistance of the beam in Example 5.1, using the stress–strain curves in Fig. 5.12.

As in Example 5.4, x may be found by equating the total compression and tension forces. For this example, the stress and strain distributions which satisfy this requirement are shown in Fig. 5.14. It can be seen that the steel stress at

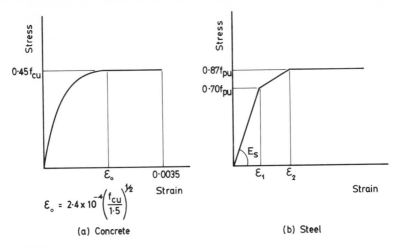

Fig. 5.12 Design stress–strain curves.

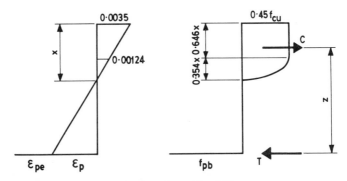

Fig. 5.13 Strain and stress distributions from Fig. 5.12.

failure of the beam is equal to the yield stress and that therefore the mode of failure is ductile.

The ultimate moment of resistance is given by

$$M_u = 1367.2 \times 0.603$$
$$= 824.4\,\text{kN m}.$$

For an applied bending moment of this amount, the total ultimate uniformly distributed load is 29.3 kN/m. The allowable characteristic imposed load w is now given by

$$1.4 \times 5.1 + 1.6w = 29.3;$$
$$\therefore w = 13.9\,\text{kN/m};$$

and the characteristic service load is thus 19.0 kN/m.

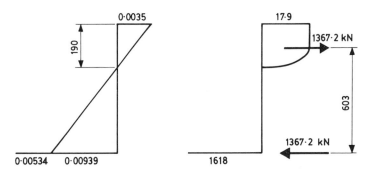

Fig. 5.14 Strain and stress distributions for Example 5.5.

In order to illustrate the different approaches using the serviceability and ultimate limit states, consider the beam designed as a Class 1 member with zero tension under the service load. For the same prestress force and eccentricity used in these examples, the allowable service load can be shown to be 15.3 kN/m. For the ultimate limit state approach, the service load is 19.0 kN/m, and so this bears out what was stated in Chapter 3, that for Class 1, and for many Class 2, members, design is usually based on the serviceability limit state.

5.8 Simplified concrete stress block

In order to simplify the calculations involved using the concrete stress block shown in Fig. 5.13, a simplified rectangular stress block is given in BS8110, shown in Fig. 5.15. This stress block gives the same total concrete force in compression as that in Fig. 5.13, and enables ultimate-strength calculations to be performed quickly by hand.

EXAMPLE 5.6   ■■

Determine the design ultimate moment of resistance of the beam in Example 5.1, using the BS8110 simplified stress block.

By equating tension and compression within the section, the neutral axis depth is found to be 211 mm, and the steel stress, from Fig. 5.12(b), is equal to the yield stress. The ultimate moment of resistance is then given by

$$M_u = 0.87 \times 1860 \times 845(700 - 0.9 \times 211/2) \times 10^{-6}$$
$$= 827.3 \, \text{kN m}.$$

The agreement between this value and that in Example 5.5 is very close.

■■

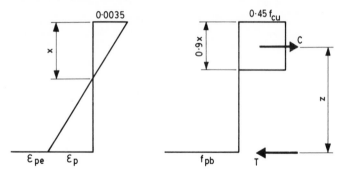

Fig. 5.15 Simplified stress block.

5.9 Code formula and design charts

As an alternative to the two methods described previously, which both use the basic principles stated in Section 5.1, the following formula for rectangular sections is given in BS8110:

$$M_u = f_{pb} A_{ps}(d - d_n),\tag{5.2}$$

where f_{pb} is the tensile stress in the prestressing steel at failure, and d_n is the depth to the centroid of the concrete stress block, taken as $0.45x$, provided that, in a flanged beam, the flange thickness is not less than $0.9x$. Values of f_{pb} and x for pretensioned and bonded post-tensioned members are given in Table 5.1. The table may be used for rectangular beams and for T-beams where the neutral axis lies within the compression flange. For T-beams where the neutral axis lies below the flange, the more rigorous methods described previously should be used. The behaviour of unbonded post-tensioned beams will be considered in Section 5.12.

Table 5.1 is shown in graphical form in Figs 5.16 and 5.17.

EXAMPLE 5.7 ■ ■

Determine the ultimate moment of resistance of the beam in Example 5.1 using Equation 5.2 and Table 5.1.

On the initial assumption that $0.9x \leqslant 200$, then

$$\frac{f_{pu} A_{ps}}{f_{cu} bd} = \frac{1860}{40} \times \frac{845}{400 \times 700}$$

$$= 0.14$$

$$\frac{f_{pe}}{f_{pu}} = \frac{880 \times 10^3}{845 \times 1860} = 0.56.$$

Table 5.1 f_{pb} and x for members with pretensioned tendons or post-tensioned tendons with effective bond.

$\dfrac{f_{pu}A_{ps}}{f_{cu}bd}$	$f_{pb}/0.87f_{pu}$			x/d		
	$f_{pe}/f_{pu} =$			$f_{pe}/f_{pu} =$		
	0.6	*0.5*	*0.4*	*0.6*	*0.5*	*0.4*
0.05	1.0	1.0	1.0	0.11	0.11	0.11
0.10	1.0	1.0	1.0	0.22	0.22	0.22
0.15	0.99	0.97	0.95	0.32	0.32	0.31
0.20	0.92	0.90	0.88	0.40	0.39	0.38
0.25	0.88	0.86	0.84	0.48	0.47	0.46
0.30	0.85	0.83	0.80	0.55	0.54	0.52
0.35	0.83	0.80	0.76	0.63	0.60	0.58
0.40	0.81	0.77	0.72	0.70	0.67	0.62
0.45	0.79	0.74	0.68	0.77	0.72	0.66
0.50	0.77	0.71	0.64	0.83	0.77	0.69

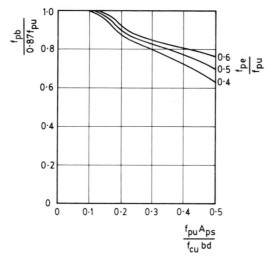

Fig. 5.16 Variation of design stress in tendons (from Table 5.1).

Thus, from Table 5.1,

$$f_{pb} = 0.97 \times 0.87 \times 1860$$
$$= 1570 \, \text{N/mm}^2$$
$$x = 0.30 \times 700$$
$$= 210 \, \text{mm} \quad (\text{i.e. } 0.9x = 189 \, \text{mm}).$$

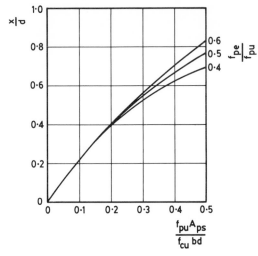

Fig. 5.17 Variation of neutral axis depth (from Table 5.2).

Therefore, from Equation 5.2

$$M_u = 1570 \times 845(700 - 0.45 \times 210) \times 10^{-6}$$
$$= 803.3 \, \text{kN m}$$

■ ■

Another method of analysis available to the designer for rectangular sections, and for T-beams where the neutral axis lies within the flange, is to use design charts. These have been prepared using stress–strain curves and a rectangular-parabolic stress block similar to those shown in Figs 5.12 and 5.13 respectively. The charts may be found in British Standard CP110:1972 Part 3; a typical chart is shown in Fig. 5.18, prepared for $f_{cu} = 40 \, \text{N/mm}^2$ and $f_{pu} = 1850 \, \text{N/mm}^2$. Although the stress–strain curves for the prestressing steel used in these charts are slightly different from that shown in Fig. 5.12(b), the charts are sufficiently accurate for design purposes.

In order to use the charts to analyse a section, the value of $100A_{ps}/bd$ is read off along the horizontal scale, and the ultimate moment of resistance of the section in terms of M_u/bd^2 is read off along the vertical scale according to the value of $100f_{pe}/f_{pu}$ for the section concerned. This quantity is a measure of the losses that have occurred and reflects the amount of strain available ε_p in the prestressing steel for use in the composite flexural behaviour of the beam. Figure 5.18 shows that for a given area of steel A_{ps}, as the value of $100f_{pe}/f_{pu}$ increases, the neutral axis depth decreases. This is because the quantity $100f_{pe}/f_{pu}$ reflects the final prestress after all losses have occurred, and as it increases, the ratio of the steel prestrain ε_{pe} to the strain in the steel at failure ε_{pb}

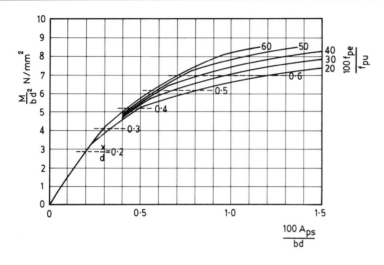

Fig. 5.18 Design chart.

increases. The amount of strain available ε_p for the steel to behave compositely with the concrete is then less, giving a larger neutral axis depth for a fixed concrete ultimate strain.

EXAMPLE 5.8   ■ ■

Determine the moment of resistance of the beam in Example 5.1, using design charts.

$100 f_{pe}/f_{pu} = 56$

$100 A_{ps}/bd = (100 \times 845)/(400 \times 700) = 0.30.$

Thus, from the design chart shown in Fig. 5.18,

$M_u = 4.2 \times 400 \times 700^2 \times 10^{-6}$

$\quad = 823.2\,\text{kN m}.$

Also, x/d is approximately 0.3, and thus the neutral axis depth is 210 mm. This is just below the flange of the beam but the moment of resistance may be taken with sufficient accuracy to be 823.2 kN m.

■ ■

Note that Fig. 5.18 is for $f_{pu} = 1850\,\text{N/mm}^2$ rather than the value of $1860\,\text{N/mm}^2$ used previously, and that the value of E_s used in the preparation of all of the design charts is $175\,\text{kN/mm}^2$, regardless of the type of steel used. Neither of these variations introduces a significant error.

5.10 Untensioned reinforcement

It is usually found that the ultimate strength of Class 1 members is satisfactory, but with some Class 2, and with most Class 3, members it is often found that the ultimate moment of resistance based on the prestressing steel alone is insufficient.

In this case, either the section size should be increased, or untensioned reinforcement added. In order to find the moment of resistance of a section with both tensioned and untensioned steel, the basic principles stated in Section 5.1 are still used, except that the relevant stress–strain curve for the untensioned reinforcement must also now be considered. Such a curve, using the character-istic yield strength of the reinforcement, f_y, is given in BS8110 and is shown in Fig. 5.19.

EXAMPLE 5.9   ■ ■

Determine the design ultimate moment of resistance of the beam section in Example 5.1, if four T10 bars are added, at the same level as the prestressing steel. Assume that $f_y = 460 \, \text{N/mm}^2$.

The same basic stress block used in Example 5.6 will be used here, except that the extra stress in the untensioned reinforcement must now be added. In this case the depth of the neutral axis is best found using a trial-and-error approach since it should not be assumed that both types of steel have yielded. The results of this analysis are shown in Table 5.2.

The strain in the prestressing steel is given by $\varepsilon_{pb} = \varepsilon_{pe} + \varepsilon_p$. The strain in the untensioned reinforcement ε_s is equal to ε_p, since both types of steel are at the same level.

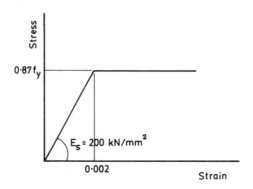

Fig. 5.19 Design stress–strain curve for reinforcement.

Table 5.2 Neutral axis depth for beam in Example 5.9.

x (mm)	ε_{pb}	ε_{st}	f_{pb} (N/mm^2)	f_{st} (N/mm^2)	T (kN)	C (kN)
200	0.0141	0.00875	1618	400	1493	1296
220	0.0130	0.00764	1602	400	1480	1426
225	0.0127	0.00739	1590	400	1469	1458

From Table 5.2, it can be seen that, with sufficient accuracy, x may be taken as 225 mm. The depth of the stress block is then 203 mm, but it is sufficiently accurate to consider only the flange as in compression. The ultimate moment of resistance is then given by:

$$M_u = 0.45 \times 40 \times 400 \times 200(700 - 200/2) \times 10^{-6}$$
$$= 864.0 \, kN \, m.$$

The ultimate moment of resistance has thus been increased from the value of 827.3 kN m found in Example 5.6. ■ ■

The full analysis of sections with untensioned reinforcement has been shown in the above example in order to illustrate the basic principles. However, in most practical cases, it is sufficiently accurate to replace the cross-sectional area of this reinforcement by an equivalent area $A_s f_y/f_{pu}$ and then to analyse the section using any of the methods described previously.

The presence of steel, either tensioned or untensioned, in the compression zone can be treated in a similar manner to the strain compatibility method described above for steel in the tension zone.

Untensioned reinforcement really becomes useful only after the concrete in a section has cracked, and particularly at the ultimate limit state. At service loads, the stress in this steel is small and may even be compressive, depending on the location within the section. The addition of untensioned reinforcement is very useful in limiting cracking and providing sufficient ultimate strength capacity soon after transfer, when the concrete is still immature. It is good practice to provide untensioned reinforcement in any region of a member where tension is likely to occur. It is of particular use at the supports of members with straight tendons, where the allowable tensile stresses, particularly at transfer, may be exceeded. This reinforcement will also resist any cracking that may occur due to accidental mishandling of structural members. This may be caused, for instance, by a simply supported beam being lifted at its centre.

5.11 Class 3 members

The three different classes of prestressed concrete member were defined in Chapter 3, and the analysis for the serviceability limit state outlined in Section 5.2 is suitable for Class 1 and 2 members, since it is assumed that the concrete remains uncracked. For Class 3 members, however, the concrete is assumed to have cracked, and the aim is to limit the crack widths to acceptable levels, depending on the degree of exposure of the member.

A procedure is given in BS8110 whereby it is assumed that the section remains uncracked, and that hypothetical tensile stresses exist corresponding to the specified maximum sizes of crack. These hypothetical stresses are shown in Table 5.3 and may be modified by the factors in Table 5.4, which take into account the reduced hypothetical tensile stresses which should be used with increased depth of the member. If additional reinforcement is contained within the tension zone, close to the face of the member, the hypothetical tensile stresses may be increased by an amount which is proportional to the cross-sectional area of the untensioned reinforcement expressed as a percentage of the cross-sectional area of the concrete in the tension zone. For 1% of reinforcement the stresses in Table 5.3 may be increased by $4.0 \, \text{N/mm}^2$ for Groups A and B and by $3.0 \, \text{N/mm}^2$ for Group C. For other percentages, the stresses may be increased in proportion, except that the total hypothetical tensile stress should not exceed $0.25 f_{cu}$.

The procedure given in BS8110 should really be regarded as a preliminary design method, suitable for estimating the section size and amount of prestressing steel. With this method, the tensile area of the concrete has a considerable effect on the steel stress, while in reality the stress in the steel is very little affected by this area of the concrete. Tests have shown that crack widths are

Table 5.3 Hypothetical tensile stresses for Class 3 members.

Group	Limiting crack width (mm)	Design stress (N/mm²) for concrete grade		
		30	40	50 and over
A Pre-tensioned tendons	0.1	—	4.1	4.8
	0.2	—	5.0	5.8
B Grouted post-tensioned tendons	0.1	3.2	4.1	4.8
	0.2	3.8	5.0	5.8
C Pre-tensioned tendons distributed	0.1	—	5.3	6.3
in the tensile zone and	0.2	—	6.3	7.3
positioned close to the tension faces of the concrete				

Table 5.4 Modification factors for Class 3 tensile stresses.

Depth of member (mm)	Factor
200 and under	1.1
400	1.0
600	0.9
800	0.8
1000 and over	0.7

more related to the steel stress than to the hypothetical tensile stress in the concrete.

Alternative design methods will be outlined in Chapter 9, but in all the design methods, the serviceability limit state of cracking must be checked. This can be carried out either by determining the crack widths directly, or by finding the stress in the steel adjacent to the tensile face of the member and checking that it is below a certain limit.

One crack-width formula (Concrete Society, 1983) is

$$w = (0.08 \times 4c f_{st}/E_s) \, \text{mm}, \tag{5.3}$$

where c is the cover to the untensioned steel and f_{st} is the stress in the untensioned steel, or increase of stress in the prestressing steel at the service load, determined using a cracked-section analysis described in Section 5.4.

As an alternative to using Equation 5.3 it is recommended (Concrete Society, 1983) that the stress f_{st} be limited to $150 \, \text{N/mm}^2$ where the whole section is in compression under the permanent load (usually the dead load of the structure plus some proportion of the imposed load). This stress may be increased to $200 \, \text{N/mm}^2$ if the full service load is of short duration. In this case any cracks which form will be temporary and will close on removal of the extra load. The allowable compressive stresses under service load, and all stresses at transfer, should be the same as for Class 2 members.

An example of the full design of a Class 3 member is given in Chapter 13.

EXAMPLE 5.10   ■ ■

For the beam in Example 5.1, determine the maximum service load which the beam can support if it is designed as a Class 3 member with a maximum crack width of 0.2 mm.

Ignoring the effect of the untensioned steel, from Table 5.3, the hypothetical tensile stress is $5.0 \, \text{N/mm}^2$. From Table 5.4, the modification factor, by interpolation, is 0.825. Thus the allowable hypothetical stress is given by

$$f_{ht} = 0.825 \times 5.0 = 4.13 \, \text{N/mm}^2$$

In order to find the maximum service load, the tensile stress at the serviceability limit state is given by

$$\frac{880 \times 10^3}{2.13 \times 10^5} + \frac{880 \times 10^3 \times 325}{35.12 \times 10^6} - \frac{M_s \times 10^6}{35.12 \times 10^6} = -4.13;$$

$$\therefore M_s = 576.1 \, \text{kN m}.$$

The corresponding uniform load is 20.5 kN/m, and this should be compared with the load of 15.3 kN/m when the beam was being treated as a Class 1 member, as shown in Section 5.7.

■ ■

The analysis of Class 3 members at the ultimate limit state is identical to that of Class 1 and 2 members.

5.12 Members with unbonded tendons

The process of grouting post-tensioned tendons after tensioning is both costly and time-consuming, and so in some structures the tendons are left unbonded. The effect of this at the serviceability limit state is very small, since in Section 5.3 it was shown that the extra stresses induced in the prestressing steel by the composite behaviour of the bonded steel and concrete are usually small

However, the behaviour at the ultimate limit state is markedly different, and the ultimate moment of resistance of an unbonded section is generally smaller than that for a similar bonded section. As the applied bending moment at a given section in such a member increases, the steel stress increases less rapidly than in a bonded section, as the increase in strain in the steel is uniform along the entire length of the member, rather than gradual along the member in line with the bending moment diagram. When the concrete crushes, therefore, the available tensile force to form the internal resisting moment is smaller than in a similar, bonded, member.

The analysis of unbonded sections at the ultimate limit state cannot be carried out based on the three basic principles stated in Section 5.1, since assumption (a) is no longer valid, that is, the strain in the steel is no longer equal to the strain in the concrete at the same level, since there is no bond between the two materials.

It is stated in BS8110 that Equation 5.2 may also be used for unbonded sections, where f_{pb} and x are given by

$$f_{pb} = f_{pe} + \frac{7000}{L/d}\left(1 - 1.7\,\frac{f_{pu}A_{ps}}{f_{cu}bd}\right), \tag{5.4}$$

$$x = 2.47 \quad A_{ps}f_{pb}/(bf_{cu}), \tag{5.5}$$

and where L is the length of the tendons between anchorages. The maximum value of f_{pb} should not be greater than $0.7f_{pu}$. The derivation of Equations 5.4 and 5.5 may be found in Rowe *et al.* (1987).

EXAMPLE 5.11　■■

The cross-section of an unbonded post-tensioned slab spanning 9 m is shown in Fig. 5.20. Determine the ultimate moment of resistance if $f_{cu} = 40\,\text{N/mm}^2$ and $f_{pu} = 1820\,\text{N/mm}^2$. Assume that total prestress losses are 25%.

For unit width of slab, $A_{ps} = 165/0.2 = 825\,\text{mm}^2/\text{m}$.

$$\begin{aligned}
\frac{f_{pu}A_{ps}}{f_{cu}bd} &= \frac{1820 \times 825}{40 \times 1000 \times 225} \\
&= 0.167, \\
L/d &= 9000/225 = 40, \\
f_{pe} &= 0.75 \times 0.7 \times 1820 \\
&= 956\,\text{N/mm}^2.
\end{aligned}$$

From Equation 5.4,

$$\begin{aligned}
f_{pb} &= 956 + (7000/40)(1 - 1.7 \times 0.167) \\
&= 1081\,\text{N/mm}^2 \quad (<0.7f_{pu}).
\end{aligned}$$

From Equation 5.5,

$$\begin{aligned}
x &= 2.47 \times (825 \times 1081)/(1000 \times 40) \\
&= 55\,\text{mm}.
\end{aligned}$$

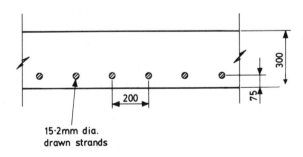

15·2mm dia.
drawn strands

Fig. 5.20

Then, using Equation 5.2,

$$M_u = 1081 \times 825(225 - 0.45 \times 55) \times 10^{-6}$$
$$= 178.6 \, kN \, m/m.$$

■ ■

Cracking in unbonded members at the ultimate limit state tends to be concentrated in a few large cracks rather than spread among many smaller cracks, as in bonded members. The addition of untensioned reinforcement will limit the width of the cracks and will also add to the ultimate strength of the member. The ultimate moment of resistance in this case may be found by replacing the area of additional steel A_s by an equivalent area of prestressing steel $A_s f_y / f_{pu}$.

A disadvantage of using unbonded tendons is that complete reliance is placed on the anchorages, so that in the event of failure of the tendons the anchorages must withstand the effect of the sudden release of strain energy stored in the tendons. With bonded tendons, the release of strain energy is also absorbed by the concrete surrounding the tendons. The lack of bond will be an advantage, however, if ever the tendons need to be re- or de-tensioned. As mentioned in Chapter 1, increasing concern is being expressed over the demolition of prestressed concrete structures which have reached the end of their useful life. If the tendons are unbonded, the force in the tendons may be transferred to a jack, depending on the type of anchorage, and then gradually reduced to zero by releasing the jack hydraulic pressure. With bonded tendons, a considerable force will still be locked in the tendons due to the bond between the steel and concrete. However this may be advantageous during demolition, since the tendons can be cut into small lengths, with each section now behaving as a pretensioned tendon.

Much of the application of prestressing to slab construction is carried out using unbonded tendons, since the large number of tendons in slabs makes grouting an expensive operation. These are often greased, wrapped in tape and cast, untensioned, into the concrete slab. The grease not only serves to destroy any bond between steel and concrete, but it also helps to reduce friction when the tendons are tensioned. Information on the protection of unbonded tendons may be found in Fédération Internationale de la Précontrainte (1986).

References

Concrete Society (1983) *Partial Prestressing*, Technical Report No. 23, London.
Fédération Internationale de La Précontrainte (1986). *Corrosion Protection of Unbonded Tendons*, London.
Mosley, W. H. and Bungey, J. H. (1987) *Reinforced Concrete Design*, Macmillan, London.
Rowe, R. E. *et al.* (1987) *Handbook to British Standard BS8110: 1985 Structural Use of Concrete*, Viewpoint, London.

Chapter 6

DEFLECTIONS

6.1 Limits to deflections

The importance of the serviceability limit state of deflection was described in general terms in Chapter 3. The effect of deflections in particular structures varies according to the use of the structure. For bridges, excessive deflections may lead to the creation of pools of water on road surfaces, a problem which can also occur where deflections of roof beams in buildings are large. If this ponding becomes too severe, an unacceptable extra dead load may be placed on the beams. Large deflections of floors may cause the cracking of partitions and windows.

It is recommended in BS8110 that for structures where the sag of a member would be noticeable, this be limited to $L/250$, where L is the span of a beam or the length of a cantilever. For structures where partitions, cladding and finishes have not been specifically designed to allow for movement of the surrounding structure, it is recommended that deflections do not exceed $L/500$, or 20 mm, whichever is less, for brittle materials, and $L/350$, or 20 mm, whichever is less, for non-brittle materials. These values may also be taken to apply to the initial upwards camber for prestressed concrete members.

The deflections of concrete structures cannot be predicted with a high degree of accuracy, since there are many non-linear factors involved. Concrete itself does not have a linear stress–strain curve, and the load–deflection characteristics of concrete beams, reinforced or prestressed, are non-linear in general, since the stiffness changes sharply once the concrete has cracked. The methods of calculation outlined in the following sections should be regarded as giving only estimates of the deflections. For most structures, the best that can be said is that the deflections lie within certain bounds. If it is very important to know accurately the deflection for a particular structure, the only reliable method is to carry out tests on a model of the structure, using similar materials.

Prestressed concrete members differ from reinforced concrete ones with regard to deflections in that deflections under a given load can be eliminated entirely by the use of a suitable arrangement of prestressing. Another difference is that deflections in prestressed concrete members usually occur even with no applied load; this is known as *camber* and is generally an upwards deflection. The use of prestress to control deflections makes it difficult to specify span/depth ratios for initial estimation of member size, which is the practice for reinforced concrete members. Nevertheless, some rough guidelines may be given for simply supported beams. For beams carrying heavy loads, such as bridge beams, a span/depth ratio in the range of 20–26 for Class 1 members would be suitable, while for Class 2 or 3 floor or roof beams, a span/depth ratio in the range 26–30 would give a good initial estimate of section size. Span/depth ratios for prestressed concrete flat slabs are discussed in Chapter 13.

Another feature which affects the deflections of all concrete structures is the proportion of the total load which may be considered permanent. *Permanent loads* consist of the dead load of the structure plus any portion of the imposed load that is present all, or most, of the time. They are usually known much more accurately than the transient loads and have the largest effect on deflections. Typically, for office and domestic structures, 25% of the total load would be permanent, while for warehouse structures this figure would be 75%. An acceptable method of design when the permanent load proportion is high would be to ensure that the deflections under permanent loads are within the limits specified earlier, and then to accept that they may be exceeded under the temporary full service load.

As described in Chapter 2, concrete deforms both instantaneously under load, and also with time, due to creep. Thus the deflections of concrete structures should be assessed under both short- and long-term conditions.

6.2 Short-term deflections of Class 1 and 2 members

The prediction of the deflections of prestressed concrete members is more straightforward than for reinforced concrete members, since for Class 1 and 2 members, the concrete section remains uncracked and the ordinary strength-of-materials methods for finding deflections are applicable. There are several such methods, but the one which is used here is that based on the principle of virtual work.

The principle is best illustrated by means of a simple example. The beam in Fig. 6.1(a) is simply supported and is in equilibrium under a point load W. Some arbitrary deflected shape of the beam is shown in Fig. 6.1(b). This need have no relation to the true deflected shape of the beam under the load W; all that is required is that the displacements at every point are small and geometrically compatible with the curvature along the beam.

The principle of virtual work states that the work done by the external applied load W moving through the displacement given by the arbitrary deflected shape is equal to the internal work done along the beam during that displacement. This work is usually considered as that due to bending only. Thus

$$W\delta_0 = \int_0^L M(x)\mathrm{d}\theta.$$

where $M(x)$ is the bending moment at a section x induced by the applied load, and θ is the rotation of the member at that section due to the arbitrary

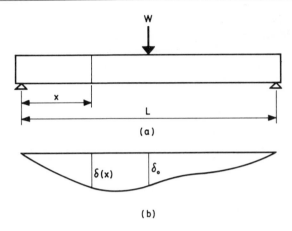

Fig. 6.1 The virtual work principle.

displacement. The way in which this principle is used to find the deflection of a structure is to apply a unit load at the point where the deflection is required, and in the same direction as the required displacement. The arbitrary deflected shape is then taken as the true deflected shape of the structure. The virtual work equation can now be written as:

$$1 \times \delta_0 = \int_0^L M'(x)\,d\theta, \tag{6.1}$$

where $M'(x)$ is the bending moment at a section x under the action of the unit point load.

The rotation $\Delta\theta$ over any small length of the beam Δx under the applied load W is given by

$$\Delta\theta = [M(x)/EI]\Delta x$$

where EI is the flexural stiffness of the beam. Thus Equation 6.1 becomes

$$\delta_0 = \int_0^L [M'(x)M(x)/EI]\,dx.$$

The integration of the two bending moment diagrams is best carried out numerically, using Simpson's rule.

In order to determine the deflections of simply supported members under prestress force only, use is made of the fact that the moment in the member at any section x is equal to $Pe(x)$ where $e(x)$ is the eccentricity at that section. The prestress moment diagram is thus proportional to the area between the member centroid and the location of the resultant prestressing force, as shown in Fig. 6.2(b) for the beam shown in Fig. 6.2(a).

With statically indeterminate prestressed concrete members, the location of the resultant prestress force is not necessarily coincident with the centroid of the

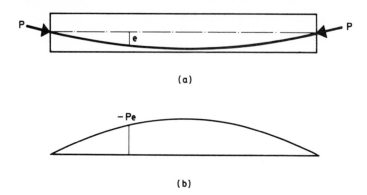

(a)

(b)

Fig. 6.2 Prestress moment diagram.

tendons and the prestress moment diagram cannot be determined as described above. Nevertheless, once the moment diagram due to the prestressing force has been determined (using the methods described in Chapter 11), the virtual work principle as outlined above can still be used to find the deflection at any point.

EXAMPLE 6.1   ■ ■

Determine the midspan deflections of the beam shown in Fig. 6.3: (i) at transfer with an initial prestress force of 6800 kN; (ii) under an imposed load of 30 kN/m when the prestress force has been reduced to 4500 kN.

$$\text{Beam self weight} = 11.26\,\text{kN/m},$$
$$\text{Total service load} = 11.26 + 30$$
$$= 41.26\,\text{kN/m}.$$

The bending moment distributions M_i and M_s are shown in Fig. 6.4(a) and Fig. 6.4(b) respectively. Figure 6.4(c) shows that due to the prestress force alone, M_p, and Fig. 6.4(d) that due to a unit vertical load at the centre of the beam, M'.

(i) At transfer, the midspan deflection is given by

$$\delta_i = \int_0^L [M'(M_i + M_p)/EI]\,\mathrm{d}x.$$

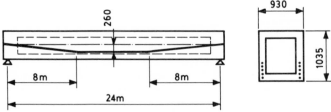

Fig. 6.3

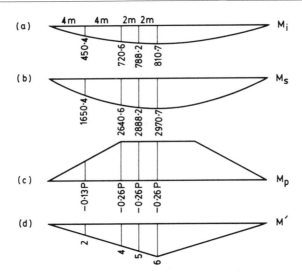

Fig. 6.4 Moments for beam in Example 6.1 (kN m).

Thus,

$$\delta_i = (2 \times 8/6EI)[4(450.4 - 0.13P)(2) + (720.6 - 0.26P)(4)]$$
$$+ (2 \times 4/6EI)[(720.6 - 0.26P)(4) + 4(788.2 - 0.26P)(5)$$
$$+ (810.7 - 0.26P)(6)].$$

With

$$P = 6800 \, \text{kN},$$

$$\delta_i = -59\,795/EI.$$

For the section used, $I = 0.06396 \, \text{m}^4$, and the short-term value of E can be taken as $28 \times 10^6 \, \text{kN/m}^2$.

$$\therefore \delta_i = -59\,795/(28 \times 10^6 \times 0.06396)$$
$$= -0.0334 \, \text{m}.$$

That is, the deflection is 33.4 mm upwards, representing a camber of 1 in 719, which is satisfactory.

(ii) At the service load, the maximum deflection is given by

$$\delta_s = \int_0^L [M'(M_s + M_p)/EI] \mathrm{d}x.$$

Thus,

$$\delta_s = (2 \times 8/6EI)[4(1650.4 - 0.13P)(2) + (2640.6 - 0.26P)(4)]$$
$$+ (2 \times 4/6EI)[(2640.6 - 0.26P)(4) + 4(2888.2 - 0.26P)(5)$$
$$+ (2970.7 - 0.26P)(6)].$$

With

$$P = 4500 \, kN,$$

$$\delta_s = 106\,482/EI$$

$$\therefore \delta_s = \frac{106\,482}{28 \times 10^6 \times 0.06396}$$

$$= 0.0595 \, m.$$

That is, the maximum service load deflection is 59.5 mm, or 1 in 404. This is less than the maximum deflection of 1 in 350 for partitions made of brittle materials and is thus satisfactory. If the beam in Example 6.1 were treated simply as an elastic beam under a uniformly distributed load, then the maximum deflection would be given by

$$\delta_s = \frac{5}{384} \times \frac{41.26 \times 24^4}{28 \times 10^6 \times 0.06396}$$

$$= 0.0995 \, m.$$

The maximum deflection would thus be 99.5 mm, or 1 in 241, just above the maximum allowable deflection for partitions of non-brittle material. The action of the prestressing force has been to reduce this deflection, giving acceptable service load behaviour.

■ ■

6.3 Deflections of Class 3 members

The ordinary strength-of-materials approach to the calculation of deflections in Class 1 and 2 members may be used for Class 3 members, provided that the tensile stresses under the permanent load do not exceed those given in Table 3.3. In other cases, it must be assumed that the section is cracked and a more rigorous analysis method should be used.

The general relationship between curvature $1/r$ at a point x along a member and the corresponding deflection y is given by

$$1/r = d^2 y/dx^2. \tag{6.2}$$

In order to find the deflection at any point in a member, Equation 6.2 must be integrated twice, and this is usually carried out numerically.

It is recommended in BS8110 that, in determining the curvature to use in Equation 6.2, it should be assumed that the section is both cracked and uncracked, and the larger of the two curvatures used.

The uncracked curvature is given by

$$1/r_u = M_x/E_c I_g, \tag{6.3}$$

where M_x is the bending moment at any section x and $E_c I_g$ is the flexural stiffness based on the gross cross-section of the member.

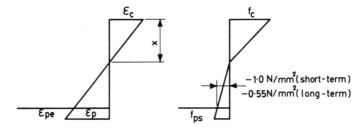

Fig. 6.5 Strain and stress distributions for cracked curvature analysis.

The cracked curvature may be found by considering the strain and stress distributions shown in Fig. 6.5. The cracked-section analysis described in Chapter 5 assumed that all of the concrete below the neutral axis is cracked. However, in practice this area of the section is not completely cracked and contributes to the stiffness of the member. The tensile stress in the concrete at the level of the prestressing steel is assumed to be 1.00 N/mm^2 instantaneously, reducing to 0.55 N/mm^2 in the long term. Once the neutral axis depth has been found from a cracked-section analysis, the cracked curvature is given by

$$1/r_c = \varepsilon_c/x.$$

A simplified method of finding the maximum deflection of concrete members is outlined in BS8110 and is suitable for Class 3 members with low percentages of prestressing steel. In this case, the maximum deflection y_{max} is given by

$$y_{max} = KL^2/r_b \tag{6.4}$$

where L is the effective span, $1/r_b$ is the curvature at midspan, or at the support

Table 6.1 Coefficient K for use in Equation 6.4

Type of loading	Bending moment diagram	K
		0.104
		$\dfrac{3-4a^2}{48(1-a)}$
		0.25
		0.125

for a cantilever, and K is a constant which depends on the shape of the bending moment diagram. The values of K for some common bending moment diagrams are shown in Table 6.1. However, the deflections for complex loading cases should not be obtained by superposition of values of K for simpler load cases. In this case, a value of K appropriate to the actual load should be used.

EXAMPLE 6.2 ■ ■

The beam shown in Fig. 6.6 has a prestress force after losses of 100 kN and supports a uniform imposed load of 24 kN/m, including its own weight. Determine the maximum short-term deflection, relative to the deflected position at transfer, if the beam has been designed as a Class 3 member, (i) using Equation 6.2, (ii) using Equation 6.4.

(i)

$$f_{pe} = 100 \times 10^3/100 = 1000 \, \text{N/mm}^2$$
$$\therefore \varepsilon_{pe} = 1000/(195 \times 10^3) = 0.00513.$$

The radius of curvature of the cracked section is found by considering equilibrium within the section, using the stress–strain curves shown in Figs 3.3 and 3.4. The strain and stress distributions within the section are based on Fig. 6.5 and are determined using a trial-and-error approach. If a programmable calculator or microcomputer is employed, the solution may be obtained very quickly. The calculations may be simplified, without introducing an unacceptable error, by ignoring the small area of concrete in tension below the level of the steel. The curvatures along the beam based on both the cracked and uncracked section are shown in Table 6.2.

The distribution of curvature to be used in finding the deflections is shown in Fig. 6.7. The change in slope between the left-hand support and the quarter-span and midspan sections are determined using Simpson's rule and are given respectively by

$$\left[\frac{dy}{dx}\right]_0^{1.875} = (1875/6)(4 \times 1.50 + 4.85) \times 10^{-7}$$
$$= 0.34 \times 10^{-3}$$

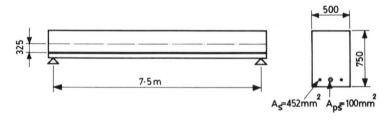

Fig. 6.6

Table 6.2 Curvatures for beam in Example 6.1.

Distance	$L/8$	$L/4$	$3L/8$	$L/2$
$(M_s - M_i)(\text{k N m})$	73.8	126.5	158.2	168.7
Cracked $1/r_c(\text{mm}^{-1} \times 10^{-7})$	0.74	4.85	10.31	12.40
Uncracked $1/r_u(\text{mm}^{-1} \times 10^{-7})$	1.50	2.57	3.21	3.43

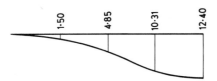

Fig. 6.7 Distribution of curvature for beam in Example 6.2.

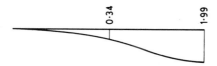

Fig. 6.8 Distribution of slope for beam in Example 6.2.

and

$$\left[\frac{dy}{dx}\right]_0^{3.75} = (3750/6)(4 \times 4.85 + 12.40) \times 10^{-7}$$
$$= 1.99 \times 10^{-3}.$$

The distribution of slope along the beam relative to the left-hand support is thus as shown in Fig. 6.8. The difference in displacement between the left-hand support and the midspan section, and thus the central deflection, is then given by:

$$[y]_0^{3.75} = y_{max} = (3750/6)(4 \times 0.34 + 1.99) \times 10^{-3}$$
$$= 2.1 \text{ mm.}$$

(ii) From Table 6.1, $K = 0.104$.
From Table 6.2, the cracked curvature at midspan is $12.4 \times 10^{-7} \text{mm}^{-1}$.
Thus

$$y_{max} = 0.104 \times 7500^2 \times 12.4 \times 10^{-7}$$
$$= 7.3 \text{ mm.}$$

An alternative to the method of determining deflections given in BS8110 is that specified by the American code ACI 318-77. This uses an effective second moment of area given by

$$I_e = \left(\frac{M_{cr}}{M_{max}}\right)^3 I_g + \left[1 - \left(\frac{M_{cr}}{M_{max}}\right)^3\right] I_{cr}, \tag{6.5}$$

where I_g and I_{cr} are the second moments of area of the gross and cracked sections respectively, M_{cr} is the bending moment to cause cracking at the tension face and M_{max} is the maximum bending moment in the member. The value of I_e determined from Equation 6.5 may then be used in conjunction with Equation 6.3.

6.4 Load balancing

The fact that an eccentric prestress force gives rise to vertical deflections in a prestressed concrete member leads to a method of design known as *load balancing*. By suitable adjustment of the prestress force and eccentricity, the deflection of a member can be made to be zero under all, or some proportion, of the service load. In this case the stresses in the member are purely axial. The extra deflections under any unbalanced load may be determined using any of the usual strength-of-materials methods.

There is much room for judgement on the part of the designer as to what proportion of the total service load should be balanced. If the total load is balanced, then an unacceptable camber may result at transfer. If only the dead load is balanced, then service load deflections may be excessive. One common criterion used in design is to balance the permanent load on the structure, as defined in Section 6.1.

Theoretically, a uniform applied load can only be balanced by a continuously draped tendon, while a concentrated load can only be balanced by a sharp change of curvature. In practice, however, any given load can be balanced approximately by either type of tendon, or a combination of both.

EXAMPLE 6.3   ■ ■

For the beam shown in Fig. 6.9, determine the prestress force required to balance a total applied load of 20 kN/m. Determine, also, the stress distribution at midspan.

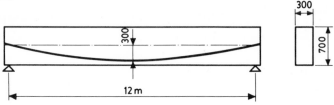

Fig. 6.9

From Chapter 1, the uniform lateral load from a parabolic tendon is given by

$$w = 8Pe/L^2$$
$$\therefore P = wL^2/8e.$$

Thus the prestress force required to balance a uniform load of 20 kN/m is given by

$$P = (20 \times 12^2)/(8 \times 0.3)$$
$$= 1200\,\text{kN}.$$

Section properties:

$$A_c = 2.1 \times 10^5\,\text{mm}^2$$
$$Z_b = Z_t = 24.5 \times 10^6\,\text{mm}^3$$

At midspan,

$$M_s = (20 \times 12^2)/8$$
$$= 360\,\text{kN m}.$$
$$f_t = \frac{1200 \times 10^3}{2.1 \times 10^5} - \frac{1200 \times 10^3 \times 300}{24.5 \times 10^6} + \frac{360 \times 10^6}{24.5 \times 10^6}$$
$$= 5.71 - 14.69 + 14.69$$
$$= 5.71\,\text{N/mm}^2.$$
$$f_b = 5.71 + 14.69 - 14.69$$
$$= 5.71\,\text{N/mm}^2.$$

That is, the section is in a state of uniform compression, as expected. ■ ■

6.5 Long-term deflections

The deflections for prestressed concrete members determined so far have been those in the short term and are caused by elastic deformation of the concrete in response to loading. However, long-term shrinkage and creep movements will cause the deflections of prestressed concrete members to increase with time.

The effects of creep may be estimated by using a method given in BS8110 whereby an effective modulus of elasticity, $E_{c_{eff}}$, is given by

$$E_{c_{eff}} = E_{c_t}/(1 + \phi) \tag{6.6}$$

where E_{c_t} is the instantaneous modulus of elasticity at the age considered and ϕ is the creep coefficient defined in Chapter 2. The value of E_{c_t} may be estimated from Equation 6.7, where $E_{c_{28}}$ and $f_{cu_{28}}$ are the modulus of elasticity and characteristic strength of the concrete, respectively, at 28 days and f_{cu_t} is the characteristic strength at the given age, which may be found from Table 6.3.

$$E_{c_t} = E_{c_{28}}\left(0.4 + \frac{0.6}{f_{cu_{28}}}f_{cu_t}\right). \tag{6.7}$$

Table 6.3 Development of concrete strength with age.

Grade	Characteristic strength f_{cu} (N/mm^2)	Cube strength (N/mm^2) at an age of:				
		7 days	2 months	3 months	6 months	1 year
20	20.0	13.5	22	23	24	25
25	25.0	16.5	27.5	29	30	31
30	30.0	20	33	35	36	37
40	40.0	28	44	45.5	47.5	50
50	50.0	36	54	55.5	57.5	60

When a simply supported beam shrinks, it generally does not do so uniformly across the section, since the concentration of steel on the tension face is usually greater than on the compression face. The restraint provided by this steel gives rise to an extra component of deflection at midspan. A method of determining this deflection is given in Part 2 of BS8110, although these shrinkage effects are usually small and are often ignored.

All of the foregoing methods for estimating deflections are applicable only when the full service load is permanent. In the usual case where only a proportion of the service load is permanent, the long-term curvature of a section may be found using the following procedure:

(a) Determine the short-term curvature under the permanent load.
(b) Determine the short-term curvature under the total load.
(c) Determine the long-term curvature under the permanent load.

Then,

Total long-term curvature = Long-term curvature under permanent load
                          + short-term curvature under non-permanent load
                          = Curvature (c) + curvature (b) − curvature (a).

EXAMPLE 6.4   ■ ■

Determine the long-term deflection of the beam shown in Example 6.1 if two-thirds of the total service load is permanent. Assume $f_{cu_{28}} = 40\,N/mm^2$.

From Table 6.3, the characteristic strength in the long term (more than 1 year) is $50\,N/mm^2$, and from Table 2.1, E_{28} is $30\,kN/mm^2$. Thus, from Equation 6.7,

$$E_{c_\infty} = 30[0.4 + (0.6 \times 50)/40]$$
$$= 34.5\,kN/mm^2.$$

For outdoor exposure, the ambient relative humidity may be assumed to be

85%. The effective section thickness t_{eff} is given by

$$t_{eff} = \frac{2 \times 47.82 \times 10^4}{2(930 + 1035)}$$

$$= 243.4 \, mm.$$

Thus, from Fig. 2.4, ϕ is 0.9, and from Equation 6.6

$$E_{c_{eff}} = 34.5/(1 + 0.9)$$

$$= 18.2 \, kN/mm^2.$$

Following the same procedure for finding deflections as shown in Example 6.1, the short-term deflection under permanent load may be shown to be given by

$$\delta_a = 47\,070/EI = 47\,070/(28 \times 10^6 \times 0.06396)$$

$$= 0.0263 \, m.$$

The short-term deflection under full service load δ_b was found in Example 6.1 to be 0.0595 m. The long-term deflection under permanent load is given by

$$\delta_c = 47\,070/(18.2 \times 10^6 \times 0.06396)$$

$$= 0.0404 \, m.$$

The total long-term deflection is now given by

$$\delta_t = \delta_c + \delta_b - \delta_a$$

$$= 0.0404 + 0.0595 - 0.0263$$

$$= 0.0736 \, m, \text{ or } 73.6 \, mm.$$

■ ■

6.6 Load-deflection curves

Typical load–deflection curves for a beam with varying degrees of prestress are shown in Fig. 6.10, based on Abeles (1971). In each case, the ultimate strength is made the same by addition of untensioned reinforcement where necessary. Curve (a) represents a beam that is over-prestressed, and shows that failure occurs suddenly with little warning given. Curve (b) represents a Class 1 and curve (c) a Class 2 beam. The much better warning given of imminent failure by these beams is evident from the larger deflections prior to collapse. Curve (d) represents a Class 3 member, and this curve is similar to that for a reinforced concrete beam with a small percentage of steel.

In curves (b), (c) and (d), the sharp change in stiffness of the beam is clearly seen, corresponding to the point where the beam is cracked. Also shown are the different initial cambers associated with each degree of prestress.

The much smaller deflections of the Class 1 and 2 members at failure

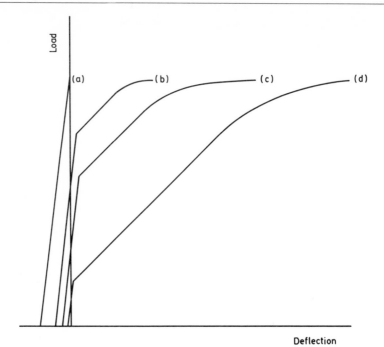

Fig. 6.10 Load–deflection curves for beam with varying prestress.

compared with Class 3 and reinforced concrete members are due to the fact that a large proportion of the strain required to achieve the large stress in the tendons at failure is locked into the member at transfer. The extra flexural strain induced is thus relatively small.

Reference

Abeles, P. W. (1971) *The Structural Engineer*, **49**, No. 2, 67–86.

Chapter 7

SHEAR

7.1 Introduction

The approach to the determination of the shear resistance of prestressed concrete members has changed from that in the earlier CP115, where the principal tensile stresses were checked at the serviceability limit state, to one where the shear resistance is checked at the ultimate limit state. Both CP110 and BS8110 embody this approach.

The shear resistance of prestressed concrete members at the ultimate limit state is very much dependent on whether or not the section in the region of greatest shear force has cracked. The actual mode of failure is different for the two cases. If the section is uncracked in flexure, then failure in shear is initiated by cracks which form in the webs of I- or T-sections once the principal tensile strength has been exceeded, Fig. 7.1(a). If the section is cracked in flexure, then failure is initiated by cracks on the tension face of the member extending into the compression zone, in a similar manner to the shear failure mode for reinforced concrete members, Fig. 7.1(b).

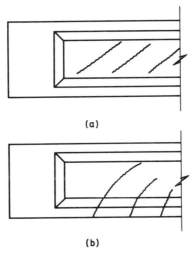

(a)

(b)

Fig. 7.1 Types of shear cracking.

7.2 Uncracked sections

A small element at the centroid of a simply supported prestressed concrete member is shown in Fig. 7.2, subjected to a compressive stress, f_{cp}, and a shear stress, f_s. The shear stress may be found from strength-of-materials theory as

$$f_s = V_{co} A\bar{y}/Ib \tag{7.1}$$

where $A\bar{y}$ is the first moment of area about the centroid of the portion above the point in the section being considered and V_{co} is the ultimate shear force at the section. A Mohr's circle analysis of the state of stress shown in Fig. 7.2 gives the value of the principal tensile stress (taken as positive here) as

$$f_{prt} = -f_{cp}/2 + \tfrac{1}{2}(f_{cp}^2 + 4f_s^2)^{1/2}. \tag{7.2}$$

Combining Equations 7.1 and 7.2 gives

$$V_{co} = (Ib/A\bar{y})(f_{prt}^2 + f_{prt}f_{cp})^{1/2}. \tag{7.3}$$

The allowable principal tensile stress is given in BS8110 by

$$f_{prt} = 0.24 f_{cu}^{1/2}, \tag{7.4}$$

where the value of f_{prt} incorporates a partial factor of safety of 1.5. The allowable principal tensile stress is related more to the direct tensile strength of the concrete, rather than to the flexural tensile strength given by Equation 2.1.

For a rectangular section, at the centroid,

$$Ib/A\bar{y} = 0.67bh$$

where b and h are the breadth and overall depth of the section respectively, and Equation 7.3 becomes

$$V_{co} = 0.67bh(f_{prt}^2 + f_{prt}f_{cp})^{1/2}. \tag{7.5}$$

In order to have the same partial factor of safety as in Equation 7.4, the stress at the centroid of the section should be taken as $0.8f_{cp}$.

Thus, Equation 7.5 becomes

$$V_{co} = 0.67bh(f_{prt}^2 + 0.8f_{prt}f_{cp})^{1/2}, \tag{7.6}$$

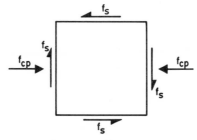

Fig. 7.2 Small element near member centroid.

Table 7.1 Values (in N/mm²) of V_{co}/bh.

f_{cp}	Concrete grade			
(N/mm^2)	30	40	50	60
2	1.30	1.45	1.60	1.70
4	1.65	1.80	1.95	2.05
6	1.90	2.10	2.20	2.35
8	2.15	2.30	2.50	2.65
10	2.35	2.55	2.70	2.85
12	2.55	2.75	2.95	3.10
14	2.70	2.95	3.15	3.30

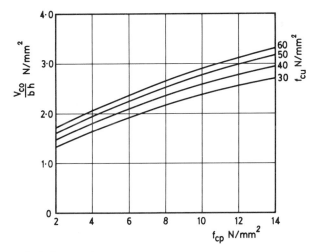

Fig. 7.3 Uncracked shear resistance.

which is the expression given in BS8110. Values of V_{co}/bh for different concrete grades and levels of prestress are given in Table 7.1 and shown graphically in Fig. 7.3, based on Equations 7.4 and 7.6.

The maximum principal tensile stress does not necessarily occur at the centroid of a section, but the above method has been found to be satisfactory. For I- and T-sections, the maximum principal tensile stress occurs at the junction of the flange and the web, but in these sections the value of $A\bar{y}$ at the junctions is greater than $0.67bh$, and so Equation 7.6 will give a reasonable approximation to the uncracked shear resistance of such sections if the centroid of the section lies within the web. However, if the centroid lies within the flange, the principal tensile stress at the junction should be limited to $0.24f_{cu}^{1/2}$. In this case, f_{cp} for use in Equation 7.2 should be 0.8 of the prestress in the concrete at the junction.

The presence of an appreciable level of vertical prestress in a member, due to draped main prestressing tendons or separate inclined or vertical tendons near the supports, would change the basic stress distribution for the small element at the centroid of the section shown in Fig. 7.2. There would now also be a vertical compressive stress, f_{cv}, and the principal tensile stress would be given by

$$f_{prt} = -(f_{cp} + f_{cv})/2 + \tfrac{1}{2}[(f_{cp} - f_{cv})^2 + 4f_s^2]^{1/2}.$$

Equation 7.5 now becomes

$$V_{co} = 0.67bh[f_{prt}^2 + 0.8f_{prt}f_{cp} + 0.8f_{prt}f_{cv} + 0.64f_{cp}f_{cv}]^{1/2}. \tag{7.7}$$

Clearly, this reduces to Equation 7.6 if f_{cv} is zero.

For pretensioned members, the critical section for shear should be taken at a distance from the face of the support equal to the height of the centroid above the soffit of the member. If this section lies within the transmission length (see Chapter 8), then the compressive stress, f_{cpx}, at any section a distance x from the end of the member is given by

$$f_{cpx} = \frac{x}{L_t}\left(2 - \frac{x}{L_t}\right)f_{cp}, \tag{7.8}$$

where f_{cp} is the full prestress at the end of the transmission length L_t. The value of L_t should not be taken as less than the depth of the member.

7.3 Cracked sections

In the case of shear failure of prestressed concrete beams which are cracked in flexure in the regions of large shear force, tests have shown that a large crack appears at a distance of half the effective depth away from the point of maximum bending moment, and then extends into the compression region of the beam, finally causing the upper fibres to crush and the beam to fail.

An empirical expression for a lower bound to the ultimate cracked shear resistance of beams, V_{cr}, has been determined as

$$V_{cr} = 0.037bd f_{cu}^{1/2} + M_{cr}V/M, \tag{7.9}$$

where d is the effective depth of the prestressing steel, M_{cr} is the cracking moment, and V and M are the shear force and bending moment, respectively, at the point where the shear crack forms. The form of Equation 7.9 shows that the cracked shear resistance comprises two components, one depending on the tensile strength of the concrete and the other on the shear force in the beam at the section where the initial cracking first extends into an inclined shear crack.

The cracking moment may be written as:

$$M_{cr} = M_0 + 0.37f_{cu}^{1/2}I/y \tag{7.10}$$

where $0.37f_{cu}^{1/2}$ is the tensile strength of the concrete and M_0 is the bending moment to produce zero stress at the extreme tension fibre, given by

$$M_0 = 0.8 f_{pt} I/y,$$

where f_{pt} is the level of prestress in the concrete at the tensile face and y is the distance of this face from the centroid of the section.

Substitution of Equation 7.10 into Equation 7.9 gives

$$V_{cr} = 0.037 bd f_{cu}^{1/2} + 0.37 f_{cu}^{1/2} I V/y M + M_0 V/M. \tag{7.11}$$

For a rectangular section, $I/y = bh^2/6$, and at a section where shear cracks occur M is approximately equal to $4Vh$, so that Equation 7.11 may be written as

$$\frac{V_{cr}}{bd} = 0.037 f_{cu}^{1/2} + 0.37 f_{cu}^{1/2} \frac{h^2}{6d} \frac{1}{4h} + \frac{M_0}{bd} \frac{V}{M}.$$

For a concrete with $f_{cu} = 50 \text{ N/mm}^2$, and with $d = h$,

$$V_{cr}/bd = 0.37 + M_0 V/bdM.$$

The tests on which Equation 7.9 was based were conducted at levels of prestress in the steel greater than $0.5 f_{pu}$. In order to develop a single expression which caters for all classes of member, Equation 7.11 is modified to

$$V_{cr} = (1 - n f_{pe}/f_{pu}) v_c bd + M_0 V/M, \tag{7.12}$$

where n is a constant and v_c is the allowable shear stress for a concrete section, and is identical to the allowable shear stress used for reinforced concrete members. When $f_{pe} = 0.6 f_{pu}$, Equation 7.12 should give

$$(1 - 0.6n) v_c = 0.37 \text{ N/mm}^2.$$

For an average value of v_c of 0.55 N/mm^2, $n = 0.55$. Thus,

$$V_{cr} = (1 - 0.55 f_{pe}/f_{pu}) v_c bd + M_0 V/M, \tag{7.13}$$

which is the expression given in BS8110 for the cracked ultimate shear resistance of a rectangular section. The value of V_{cr} should not be less than $0.1 bd f_{cu}^{1/2}$. The validity of the many assumptions made in the derivation of Equation 7.13 has been verified in tests (Reynolds, Clarke and Taylor, 1974). The cracked shear resistance determined from Equation 7.13 may be assumed to be constant for a distance equal to $d/2$ in the direction of increasing bending moment. Thus, for cantilevers, the cracked shear resistance need only be checked at a distance of $d/2$ from the face of the support.

The values of v_c are given in Table 7.2, where the total area of reinforcement $(A_s + A_{ps})$ should be used. The figures in the table have been derived from the formula

$$v_c = 0.79(100 A_s/bd)^{1/3}(400/d)^{1/4}/\gamma_m$$

on the basis of a γ_m of 1.25 and f_{cu} of 25 N/mm^2. For characteristic strengths

Table 7.2 Allowable concrete shear stress v_c (N/mm²).

$\dfrac{100A_s}{bd}$	Effective depth (mm)							
	125	150	175	200	225	250	300	$\geqslant 400$
$\leqslant 0.15$	0.45	0.43	0.41	0.40	0.39	0.38	0.36	0.34
0.25	0.53	0.51	0.49	0.47	0.46	0.45	0.43	0.40
0.50	0.67	0.64	0.62	0.60	0.58	0.56	0.54	0.50
0.75	0.77	0.73	0.71	0.68	0.66	0.65	0.62	0.57
1.00	0.84	0.81	0.78	0.75	0.73	0.71	0.68	0.63
1.50	0.97	0.92	0.89	0.86	0.83	0.81	0.78	0.72
2.00	1.06	1.02	0.98	0.95	0.92	0.89	0.86	0.80
$\geqslant 3.00$	1.22	1.16	1.12	1.08	1.05	1.02	0.98	0.91

greater than this, but less than or equal to $40\,\text{N/mm}^2$, the values in Table 7.2 may be multiplied by a factor of $(f_{cu}/25)^{1/3}$.

The depth d is from the compression face of the member to the centroid of the total steel area $(A_{ps} + A_s)$ in the tension zone. The failure of cracked prestressed concrete members is similar to that of reinforced concrete members, where the dowel action of the reinforcing bars crossing the critical shear crack increases the overall shear resistance of the member. The prestressing tendons exhibit a similar dowel action and increase the shear strength. It is interesting to note that if $f_{pe} = 0$, so that the member is no longer prestressed, this implies that $M_0 = 0$ and Equation 7.13 reduces to the expression given in BS8110 for the shear resistance of reinforced concrete members, again emphasizing the similarity in behaviour.

An important distinction between the assessment of the shear strength of uncracked and cracked sections is that, for uncracked sections, the shear resistance contributed by any inclined tendons may be taken into account. However, for cracked sections, tests have shown that inclined tendons actually decrease the shear resistance, since their influence on the concrete in the tension face is reduced, thus leading to earlier cracking. Therefore, the effect of inclined tendons on cracked sections should only be considered where it increases the effective shear force at a section.

Equation 7.13 shows that if $M < M_0$ at a section, that is it is uncracked, V_{cr} will be greater than V, the applied shear force. Thus, in uncracked regions, only V_{co} need be considered. In cracked regions, however, both V_{co} and V_{cr} should be evaluated and the lesser taken as the shear resistance of the section.

7.4 Shear reinforcement

If the shear resistance of a prestressed concrete member determined using the methods in the previous sections is insufficient, then shear reinforcement must be provided. This is usually in the form of links, similar to those used in reinforced concrete members.

It is stated in BS8110 that shear reinforcement is not required in cases where the ultimate shear force at a section is less than $0.5V_c$, where V_c is based on the lesser of V_{co} and V_{cr}, or when the member is of minor importance. Where minimum links are provided in a member, the shear resistance of these links is added to that of the member, V_c. The cross-sectional area A_{sv} of the minimum links at a section, and their spacing s_v along the member, are given by

$$A_{sv}/s_v = 0.4b/(0.87 f_{yv}), \tag{7.14}$$

where b is the breadth of the member (or of the rib in a T- or I-section) and f_{yv} is the characteristic strength of the shear reinforcement.

The total shear resistance, V_r, of a member with nominal reinforcement is given by

$$V_r = V_c + 0.4bd, \tag{7.15}$$

where d is the depth to the centroid of the tendons and longitudinal reinforcement. If the shear force at a given section exceeds the value given by Equation 7.15, then the total shear force in excess of V_c must be resisted by the shear reinforcement. The amount to be provided is then given by

$$A_{sv}/s_v = (V - V_c)/(0.87 f_{yv} d). \tag{7.16}$$

This is equivalent to the expression given in BS8110 for the shear reinforcement required in reinforced concrete members. The links used as shear reinforcement should be properly anchored and enclose all the tendons and reinforcement within the section. The shapes of links commonly used in reinforced concrete members are also suitable for use in prestressed concrete members. They should pass around a longitudinal bar or tendon with at least the same diameter as the link. The spacing of links should not exceed $0.75d$, or four times the web width for T- or I-sections. When V exceeds $1.8V_c$, the maximum spacing should be reduced to $0.5d$.

In order to prevent the crushing of webs due to the diagonal thrusts induced by shear forces, the maximum shear stress at any section should under no circumstances exceed $0.8f_{cu}^{1/2}$, or $5\,\text{N/mm}^2$, whichever is less, whether the section is cracked or uncracked. In determining this maximum stress, the reduction in web width due to ungrouted post-tensioning ducts should be considered. Even for grouted construction, only the concrete plus one-third of the duct width should be used in finding the maximum shear stress. The lateral spacing of the legs of the links across a section should not exceed d.

EXAMPLE 7.1   ■ ■

The beam shown in Fig. 7.4 supports an ultimate load, including self weight, of $85\,\text{kN/m}$ over a span of $15\,\text{m}$ and has a final prestress force of $2000\,\text{kN}$. Determine the shear reinforcement required. Assume that $f_{cu} = 40\,\text{N/mm}^2$.

Section properties: $A_c = 2.9 \times 10^5\,\text{mm}^2$

$$I = 3.54 \times 10^{10}\,\text{mm}^4.$$

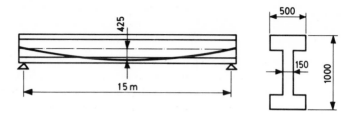

Fig. 7.4

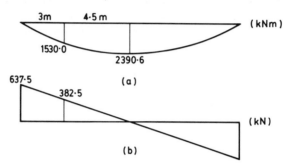

Fig. 7.5 Bending moment and shear force diagrams for beam in Example 7.1.

The ultimate bending moment and shear force diagrams are shown in Figs 7.5(a) and (b), respectively.

The maximum allowable shear force in the section is given by
$$V_{max} = 0.8 \times 40^{1/2} \times 150 \times 1000 \times 10^{-3}$$
$$= 758.9 \, kN.$$

For the uncracked regions of the beam, since the centroid of the section lies within the web, the shear resistance is given by Equation 7.6 with f_{cp} taken at the centroid

$$f_{cp} = 2000 \times 10^3/(2.9 \times 10^5)$$
$$= 6.90 \, N/mm^2.$$

Thus, from Table 7.1, $V_{co}/bh = 2.19 \, N/mm^2$ and

$$V_{co} = 2.19 \times 1000 \times 150 \times 10^{-3}$$
$$= 328.5 \, kN.$$

At the supports, the slope of the prestressing tendons is given by

$$\theta = \tan^{-1}(4d_r/L),$$

where d_r is the drape of the tendons.

$$\therefore \theta = \tan^{-1}(4 \times 425/15000)$$
$$= 6.47°.$$

The vertical component of the shear force at the face of the support is $2000 \sin \theta$, or $225.2 \, \text{kN}$, so that the total shear resistance at the support is $553.7 \, \text{kN}$.

The total uncracked shear resistance will decrease at points further into the beam, assuming a constant effective prestress force along its length, since the angle of inclination of the tendons, θ, will reduce. The variation of V_{co} along the length of the beam is shown in Fig. 7.6.

For the cracked regions of the beam, the shear resistance is given by Equation 7.13. The area of the prestressing tendons is $2010 \, \text{mm}^2$ and f_{pe}/f_{pu} is assumed to be constant at 0.6. At 3 m from the support, the depth of the tendons is given by

$$d = 500 + (4 \times 425/15^2) \times 3(15 - 3)$$
$$= 772 \, \text{mm}.$$

Thus,

$$100 A_s/bd = (100 \times 2010)/(150 \times 772)$$
$$= 1.74.$$

Therefore, from Table 7.2, $v_c = 0.76 \, \text{N/mm}^2$.

$$f_{pt} = \frac{P}{A} + \frac{P_{ey}}{I}$$
$$= \frac{2000 \times 10^3}{2.9 \times 10^5} + \frac{2000 \times 10^3 \times 272 \times 500}{3.54 \times 10^{10}} = 14.58 \, \text{N/mm}^2.$$
$$\therefore M_0 = (0.8 \times 14.58 \times 3.54 \times 10^{10}/500) \times 10^{-6} = 825.8 \, \text{kN m}.$$

Also,

$$V = 382.5 \, \text{kN}; \quad M = 1530.0 \, \text{kN m}.$$
$$\therefore V_{cr} = (1 - 0.55 \times 0.6) \times 0.76 \times 150 \times 772 \times 10^{-3}$$
$$+ (825.8 \times 382.5/1530.0).$$
$$= 265.4 \, \text{kN}.$$

Also,

$$V_{cr} \geqslant 0.1 \times 150 \times 772 \times 40^{1/2} \times 10^{-3} = 73.2 \, \text{kN}.$$

The cracked shear resistance must be checked at other sections along the beam and a plot of this is shown in Fig. 7.6.

It can be seen from Fig. 7.6 that everywhere along the beam, except very near the supports, $V_{cr} < V_{co}$, and that, except for a small region near the centre of the beam, $V > 0.5 \, V_c$, where V_c is the lesser of V_{co} and V_{cr}, and thus shear reinforcement is required. The nominal links required are given by

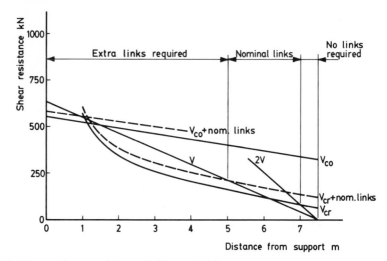

Fig. 7.6 Shear resistance of beam in Example 7.1.

Equation 7.14, i.e.

$$A_{sv}/s_v = (0.4 \times 150)/(0.87 \times 250)$$
$$= 0.276.$$

For R8 links at 350 centres, $A_{sv}/s_v = 0.287$. The total shear resistance of the beam with nominal links throughout is shown in Fig. 7.6, and it can be seen that this shear resistance is adequate for the centre 5 m of the beam, but that everywhere else more substantial reinforcement is required.

The maximum difference between the applied shear force and V_c is approximately 130 kN, and this must be resisted by shear reinforcement, with cross-sectional area and spacing given by Equation 7.16.

That is,

$$A_{sv}/s_v = (130 \times 10^3)/(0.87 \times 250 \times 772)$$
$$= 0.774.$$

Thus R 12 links at 275 mm centres would be adequate, giving $A_{sv}/s_v = 0.823$.
■ ■

References

Reynolds, G. C., Clarke, S. L. and Taylor, H. P. J. (1974) *Shear Provisions for Prestressed Concrete in The Unified Code CP110:1972.* Cement and Concrete Association Technical Report No. 42.005, London.

Chapter 8

PRESTRESSING SYSTEMS AND ANCHORAGES

8.1 Pretensioning systems

The essential features of the pretensioning of concrete members were described in Chapter 1. The anchorages used to maintain the tendons in tension until the concrete has hardened sufficiently must be reusable and the commonest system is a single barrel-and-wedge arrangement, shown in Fig. 8.1. Once the tendons have been tensioned to the required level, the jack is released and the wedges lock against the sides of the tendon and the barrel as the tendon contracts. The barrels bear directly against an anchor block which transmits the tensioning force *via* the prestressing bed to the other end of the tendon.

Once the concrete in the member has reached sufficient strength for transfer

Fig. 8.1 Pretensioning anchorage.

to take place, a prefabricated 'stool' is inserted between the anchor block and the jack. The barrel-and-wedge anchorage is relieved of its pressure by jacking the tendon to its original force, and then the loosened anchorage assembly may be removed. The jack pressure is released and the prestressing force is transferred to the concrete members along the prestressing bed.

Where deflected tendons are required in pretensioned members, the strands are initially tensioned in their original straight profiles and then deflected up (or down) at the desired locations by hydraulic jacks. They are then locked in this deflected position by a holding-down device, securely fixed to the prestressing bed. The jack used to deflect the tendons may then be removed.

An alternative method is to tension the tendons in their deflected shape, with the holding-down device in place and secured to the prestressing bed before tensioning begins. With this system, the friction between the tendons and the holding-down device must be taken into account when determining the expected extension of the tendons at the anchor block.

8.2 Post-tensioning systems

The main difference between the anchorages used in post-tensioning and pretensioning is that, in the latter, the anchorages should be reusable, but in the former the anchorages must be cast into the member and can only be used once.

There are many proprietary systems of post-tensioning anchorage available, but they fall into three main categories.

(a) Wedge anchorages

These are similar to the usual type of anchorage used in pretensioning work and may be used for tendons comprising either wires or strands. The restraining force on the tendon is provided by friction between the tendon and wedges bearing against the sides of a tapered hole in a steel plate or block. However, unlike most pretensioning applications, where each tendon is composed of a single wire or strand, in post-tensioned concrete members the tendons usually consist of many strands or wires running through the same duct. These wires or strands are usually all anchored within the same anchorage, which is more economical than providing separate anchorages for each component of the tendon.

One such barrel-and-wedge anchorage is shown in Fig. 8.2, both as a tensioning and as a non-tensioned or 'dead-end' anchorage. Note that both types incorporate helical reinforcement around the main body of the anchorage; this is to counter the tensile forces which are set up within the concrete by the large concentrated force applied to it through the anchorage. This will be discussed in more detail in Section 8.3. Each individual strand in the tensioning anchorage has its own set of wedges within a block and the force in each strand is transmitted to the concrete through a steel bearing plate.

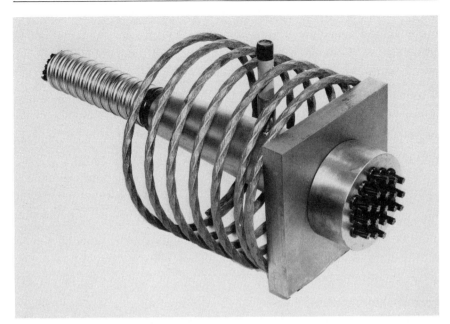

(a)

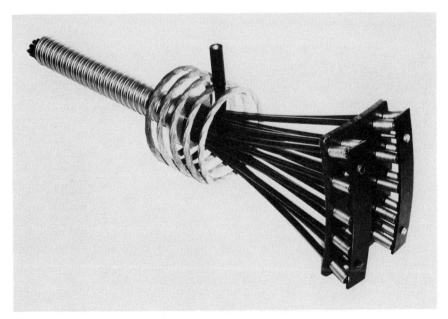

(b)

Fig. 8.2 Post-tensioning wedge anchorages: (a) tensioning anchorage; (b) dead-end anchorage (courtesy VSL International).

Tensioning is carried out using a single hydraulic jack for each tendon. All the strands in the tendon are gripped simultaneously by the jack and pulled until the desired extension is reached. The jack pressure is then released and the slight draw-in of the strands locks the wedges firmly in their seating in the block. For large tendon forces, the jacks required are very heavy and must be positioned using a crane or a hoist. Some systems which also use the wedge principle allow each strand in the tendon to be tensioned individually, so that a much smaller and more easily handled jack may be used. A 1000 tonne jack used in the tensioning of a bridge deck is shown in Fig. 8.3.

Also shown in Fig. 8.2 are the duct former leading from the anchorage, and also a grout inlet so that the duct may be injected with grout after the tendon has been tensioned. This is in order to provide protection to the tendon and to bond it to the surrounding concrete *via* the duct former. For most applications the duct formers are corrugated and made from thin galvanized steel strip and are flexible enough to enable them to be fixed in the curved profiles usually adopted. They are also rigid enough not to distort during concreting. In special applications, such as in nuclear reactor pressure vessels or offshore platforms, the tendons may be housed in steel tubes. Where contact with chemical products is likely, plastic ducts may be used. Once the tensioning and grouting are complete, the exposed portions of the anchorage are usually encased in concrete for protection.

Fig. 8.3 Post-tensioning using 1000 tonne jack.

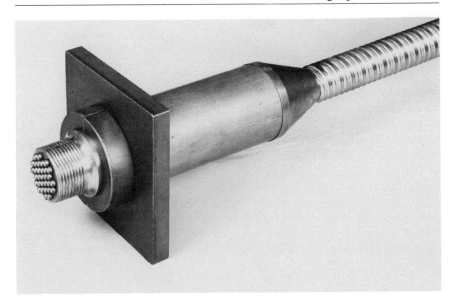

(a)

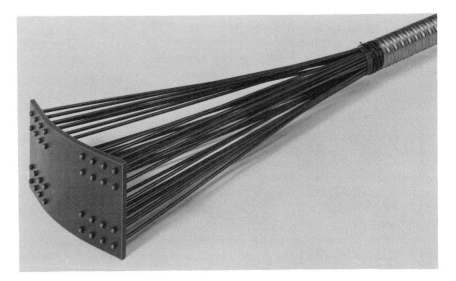

(b)

Fig. 8.4 Buttonhead anchorages: (a) tensioning anchorage; (b) dead-end anchorage (courtesy Bureau BBR Ltd).

(b) Buttonhead anchorages

For tendons comprising wires, an alternative to locking them through friction is to form a 'buttonhead' at the end of the wire and to have this shaped end bear against a plate. A typical anchorage for such a system is shown in Fig. 8.4, both as a tensioning and as a dead-end anchorage. The wires in the tensioning anchorage bear against an anchor block which is threaded to fit inside the jack, so that all the wires may be tensioned simultaneously. After tensioning, steel shims are inserted between the anchor block and the bearing plate, and these may be seen in Fig. 8.4(a).

There is negligible anchorage draw-in with this system and thus it is particularly suitable for short members. It also offers the possibility of easily adjusting or releasing the tension in the tendons at a later stage, since no protruding wires are required for gripping by the jack. One disadvantage of this system compared with a system using wedges is that the initial length of the wires must be known accurately, since the buttonheads at either end of the wire must fit snugly on the anchor block.

(c) Threaded bar anchorages

The third main category of anchorage uses a simple threaded bar, with a nut bearing against a steel plate to maintain the force in the bar. An example is shown in Fig. 8.5. If the thread is continuous, this allows greater tolerance on the length of the bar and also gives better bond with the surrounding grout inside the duct former. The grout is injected through the hole in the bearing plate after tensioning.

A ring spanner is placed over the nut and then covered by a tensioning stool which allows the nut to be tightened continuously during tensioning. The hydraulic jack bears against the stool and pulls the bar until the desired extension is achieved. By tightening the nut firmly at this stage, anchorage draw-in is virtually eliminated on release of the jack pressure.

A sequence of construction often used for multi-span bridges involves attaching tendons to the ends of those which have already been tensioned. A coupler for the buttonhead system is shown in Fig. 8.6; each system has similar arrangements.

A type of anchorage which should be mentioned briefly is that which is used for the circumferential prestressing of tanks and silos. One such anchorage is shown in Fig. 8.7. The strands pass through an anchor block from both directions and are tensioned in groups alternately. The anchor block moves back and forth within a recess in the tank wall during tensioning. Once this is complete, the anchorage is encased in concrete to give a flush surface. An alternative to this system is to have both ends of the tendon overlapping and ending in conventional anchorages. However, these anchorages would be proud of the tank surface and encased in a block-out.

Fig. 8.5 Bar anchorage (courtesy McCalls Special Products).

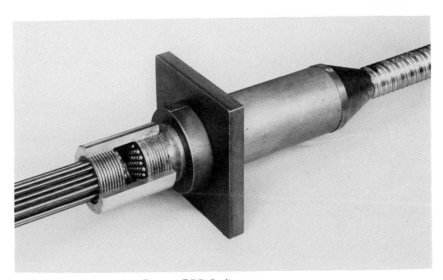

Fig. 8.6 Coupler (courtesy Bureau BBR Ltd).

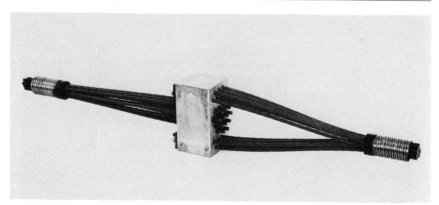

Fig. 8.7 Anchorage for circumferential prestressing (courtesy VSL International).

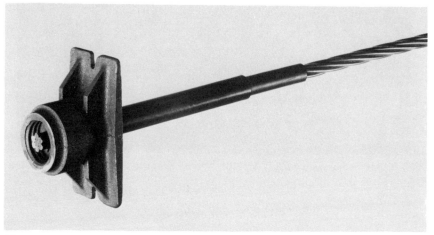

(a)

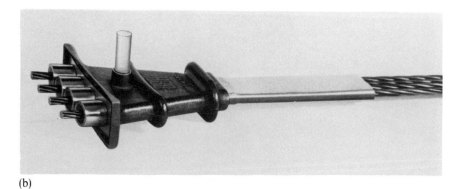

(b)

Fig. 8.8 Anchorages for slabs: (a) unbonded; (b) bonded (courtesy Bureau BBR Ltd).

Two types of anchorage commonly employed in slab construction are shown in Fig. 8.8. An anchorage for a single unbonded strand is shown in Fig. 8.8(a), while Fig. 8.8(b) shows an anchorage for a group of four bonded strands.

The anchorage details of the post-tensioning system to be used must be considered at an early stage in the design process, since the spacing of the anchorages is not only governed by the type of jack to be used but also by the bursting forces set up within the anchorage zone.

8.3 Bursting forces in anchorage zones

For post-tensioned members, the prestressing force in a tendon is applied through the anchorages as a concentrated force. By St Venant's principle, the stress distribution in a member is reasonably uniform away from the anchorage, but in the region of the anchorage itself the stress distribution within the concrete is complex. The most significant effect for design, however, is that tensile stresses are set up transverse to the axis of the member, tending to split the concrete apart, rather like a nail being driven into the end of a timber joist. Reinforcement must be provided to contain these tensile stresses.

Various theoretical and experimental studies have been carried out into the anchorage zone stresses, and the recommendations in BS8110 are based on a compromise between the methods of end-block design that have been proposed, and give results which are a reasonable approximation to the experimental results.

The end-block of a concentrically-loaded post-tensioned member of rectangular cross-section is shown in Fig. 8.9, which also shows the distributions of principal tensile and compressive stress within the end-block. The actual

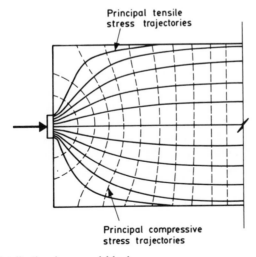

Fig. 8.9 Stress distribution in an end-block.

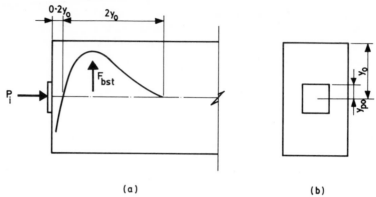

(a)   (b)

Fig. 8.10 Bursting forces.

distribution of the bursting stresses is not uniform, but varies approximately as shown in Fig. 8.10(a). It is sufficiently accurate to consider the resultant of these stresses, F_{bst}. At the serviceability limit state, F_{bst} is assumed to act in a region extending from $0.2\, y_0$ to $2\, y_0$, where y_0 is half the side of the end-block. The value of F_{bst} as a proportion of P_i, the initial tendon jacking force, may be found from Table 8.1. This shows that F_{bst} depends on the ratio of y_{po}/y_0, where y_{po} is half the side of the loaded area. Circular loaded areas should be considered as square areas of equivalent cross-sectional area.

For a given width of end-block and prestressing force, the bursting force increases as the size of the loaded area decreases, reflecting the greater splitting effect of the more concentrated force. For rectangular anchorages or end-blocks, the bursting forces should be considered in each of the principal directions based on the relevant values of y_{po}/y_0.

For post-tensioned members which are grouted after tensioning, the maximum force applied to the member is the initial jacking force, and design is based on the serviceability limit state. The bursting force should thus be resisted by reinforcement in the form of spirals or closed links, uniformly distributed throughout the end-block and with a stress in them of $200\ \mathrm{N/mm^2}$. An adequate factor of safety for the ultimate limit state is ensured partly by the fact that, by the time the tendons are grouted, the initial losses have occurred and the force applied through the anchorage is less than P_i. Also, once grouting has been completed, in the event of failure of the anchorage, some force will still be maintained in the tendon by bond to the surrounding concrete.

For post-tensioned members which are left unbonded after tensioning, in the

Table 8.1 Bursting forces in end-blocks.

y_{po}/y_0	0.2	0.3	0.4	0.5	0.6	0.7
F_{bst}/P_i	0.23	0.23	0.20	0.17	0.14	0.11

event of failure of the anchorage, all of the tendon force would be lost, and to ensure adequate safety at the ultimate limit state, the bursting force, F_{bst}, should be determined from Table 8.1 on the basis of the characteristic tendon force. No partial factor of safety need be applied to this force, since it is assumed that jacking is carried out under carefully controlled conditions. The stress in the reinforcement to contain the bursting forces should be assumed to be its design strength of $0.87f_y$.

Where an end-block contains several anchorages, it should be divided into a series of symmetrically loaded prisms and then each prism treated as a separate end-block, as described above. Additional reinforcement should be provided around the whole group of anchorages to maintain overall equilibrium. More information on the treatment of end-block design may be found in Abeles and Bardhan-Roy (1981) and in the guide of the Construction Industry Research and Information Association (1976).

EXAMPLE 8.1 ■■

The beam end shown in Fig. 8.11 has six anchorages with 75 mm square bearing plates and a jacking force of 500 kN applied to each. Determine the reinforcement required to contain the bursting forces if f_y for the reinforcement is 460 N/mm².

The overall end-block can be divided into prisms, each with an end face 250×150 mm.
For the vertical side,

$$y_{po}/y_0 = 37.5/125 = 0.3.$$

For the horizontal side,

$$y_{po}/y_0 = 37.5/75 = 0.5.$$

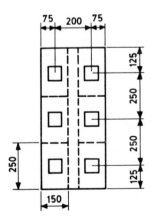

Fig. 8.11

From Table 8.1, F_{bst} in the horizontal direction is 0.23 $P_i = 115$ kN, to be resisted over a length of 250 mm from the end face. In the vertical direction, F_{bst} is 0.17 $P_i = 85$ kN, to be resisted over a length of 150 mm from the end face.

The maximum force to be resisted is thus 115 kN, and the area of reinforcement required A_{sv} based on a stress of 200 N/mm² is given by

$$A_{sv} = 115 \times 10^3/200 = 575 \, \text{mm}^2.$$

Thus four T10 closed links are required around each anchorage, giving a total area of 628 mm².

The six individual prisms may be considered as equivalent to one single prism of dimensions 750 × 350 mm. The equivalent size of loaded area is given by $(6 \times 75^2)^{1/2} = 184$ mm square.

Thus, for the vertical side,

$$y_{po}/y_0 = 92/375 = 0.25.$$

For the horizontal side,

$$y_{po}/y_0 = 92/175 = 0.53.$$

From Table 8.1, the maximum force to be resisted

$$= 0.23 \times 6 \times 500$$
$$= 690 \, \text{kN}$$

over a length of 750 mm from the end face, and the area of reinforcement required is given by

$$A_{sv} = 690 \times 10^3/200 = 3450 \, \text{mm}^2.$$

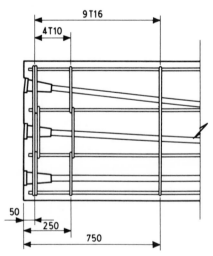

Fig. 8.12 Detailing for beam in Example 8.1.

This is provided by nine T16 links giving a total cross-sectional area of 3619 mm². The practical detailing of the end block is shown in Fig. 8.12.

■ ■

8.4 Transmission lengths in pretensioned members

Once the tendons in a pretensioned member have been cut, the force in them, which was initially maintained by the anchorages at the end of the pretensioning bed, is transferred suddenly to the ends of the concrete member. However, since there is no anchorage at the end of the member, as in the case of post-tensioning, there can be no force in the tendon there. Further along the tendon, the bond between the steel and the concrete enables the force in the tendons to build up, until at some distance from the end of the member a point is reached where the force in the tendons equals the initial prestress force. This distance is known as the *transmission length*. A typical variation of prestress force along a pretensioned member is shown in Fig. 8.13. There are many factors which affect the transmission length; the transmission lengths for wires have been found to vary between 50 and 160 diameters. An investigation into these various factors is described in Base (1958).

At the end of the member, the bond between steel and concrete tends to break down and the tendons slip relative to the concrete, thus setting up a friction force between the two materials. The force in the tendons near the ends of the member decreases after their release, and their diameters increase slightly due to the Poisson's ratio effect, thus increasing the frictional resistance. If the tendons are wires which have been mechanically crimped, this will provide better frictional resistance and thus reduce the transmission length; this is important in short members such as railway sleepers. Strands have been found to give better frictional resistance than smooth wires of equivalent area.

Further away from the free end of the member, the Poisson's ratio effect is reduced, since the reduction in tendon force is smaller, and the force in the tendons is built up primarily by bond between the steel and concrete. The bond

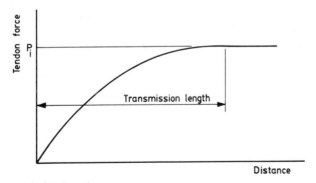

Fig. 8.13 Transmission length.

Table 8.2 Transmission length coefficient K_t.

Type of Tendon	K_t
Plain or indented wire (inc. crimped wire of small wave ht)	600
Crimped wire with wave height $\geqslant 0.15$ dia.	400
7-wire standard or super strand	240
7-wire drawn strand	360

forces developed are dependent on the concrete cube strength, being larger for high cube strengths, and also on the degree of compaction of the concrete. This is often better at the bottom of horizontally-cast members than at the top. Also, for a given area of cross-section, several small tendons have larger surface area than one larger tendon, and thus the bond forces developed will be greater, the same consideration which applies to local bond stresses in a reinforced concrete member.

Transmission lengths should be determined as far as possible from factory or site conditions, but in the absence of such data the following formula for the transmission length is given in BS8110, which is valid for initial prestress levels in the tendons of up to $0.75 f_{pu}$:

$$L_t = \frac{K_t}{f_{ci}^{1/2}} \times (\text{diameter}) \tag{8.1}$$

where f_{ci} is the cube strength at transfer and K_t is a coefficient for the type of steel used. Values of K_t are summarized in Table 8.2.

Since the bond between the steel and concrete is so important in developing the prestress force in a pretensioned member, it is important to limit the cracking which may occur near the ends of these members, both at the serviceability and ultimate limit states. This cracking would cause very high increases in bond stress adjacent to the cracks and could lead to bond slip of the steel. Provision of additional untensioned reinforcement should limit the cracking to acceptable levels.

The reduced prestress force near the ends of pretensioned members must be allowed for in the checking of stresses at various sections along the member. It is usually desirable to have a reduced eccentricity near the ends of simply supported beams, since the bending moments are lower in these regions. In post-tensioned members that is easily achieved, but in pretensioned members, although the eccentricity near the ends may be reduced by deflecting the tendons, this is an expensive procedure and an alternative is deliberately to eliminate the bond between the steel and concrete over a given length by greasing the tendons, or providing them with sleeves which allow the tendons to

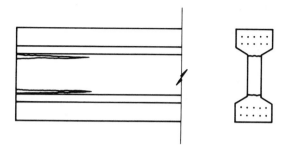

Fig. 8.14 Cracking in a pretensioned member.

move freely within them. In this case, the transmission length must be regarded as starting where the de-bonding procedure has finished.

The prestressing force in pretensioned members is applied gradually along the transmission length, and the bursting forces associated with post-tensioned members do not usually arise. However, where the pretensioning tendons are placed in two widely separated groups, as shown in Fig. 8.14, horizontal cracks may develop, and these should be restrained by the provision of link reinforcement placed around both groups of tendons.

References

Abeles, P. W. and Bardhan-Roy, B. K. (1981) *Prestressed Concrete Designer's Handbook*, Viewpoint, Slough.

Base, G. D. (1958) *An Investigation of the Transmission Length in Pretensioned Concrete*. Cement and Concrete Association Publication 41.005, London.

Construction Industry Research and Information Association (1976) *A Guide to the Design of Anchor Blocks for Post-Tensioned Prestressed Concrete Members*. Guide No. 1, London.

Chapter 9

DESIGN OF MEMBERS

9.1 Introduction

In Chapter 5, the analysis of sections under given prestress force and applied load was considered. The general problem, however, is that of design – given a structure with overall geometry and applied load, what size of member is required and what are the details of the prestressing force and tendon profile required? A trial-and-error approach could be used, the test being whether the stresses at all sections of the member are satisfactory under all possible load conditions. This might prove to be a very lengthy process, however, and a systematic approach would clearly be advantageous.

The design of Class 1 and 2 members will be illustrated in the following sections, primarily through the use of an example of a prestressed concrete bridge deck slab. It is recognized that the solid section used in the examples is not the most economical section (see Section 9.9) but it serves to illustrate the basic design principles and to introduce the idea of a prestressed concrete slab. The design procedure for Class 3 members is slightly different and will be shown in Section 9.8. Flow charts for the design procedures for Class 1, 2 and 3 members are shown in Section 9.10.

9.2 Basic inequalities

As a starting point in the design process, consider a simply supported beam carrying a uniform load, as shown in Fig. 9.1.

If the initial prestress force and eccentricity at midspan are P_i and e respectively, then the stresses at the top and bottom fibres of the beam at midspan, at transfer and under the service load, may be described by the four equations shown below:

Transfer:

$$f_t = \frac{\alpha P_i}{A_c} - \frac{\alpha P_i e}{Z_t} + \frac{M_i}{Z_t} \tag{9.1a}$$

$$f_b = \frac{\alpha P_i}{A_c} + \frac{\alpha P_i e}{Z_b} - \frac{M_i}{Z_b} \tag{9.1b}$$

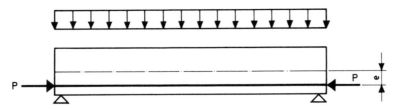

Fig. 9.1 Simply supported prestressed concrete beam.

Service load:

$$f_t = \frac{\beta P_i}{A_c} - \frac{\beta P_i e}{Z_t} + \frac{M_s}{Z_t} \tag{9.1c}$$

$$f_b = \frac{\beta P_i}{A_c} + \frac{\beta P_i e}{Z_b} - \frac{M_s}{Z_b} \tag{9.1d}$$

where Z_t and Z_b are the elastic section moduli of the section for the top and bottom fibres respectively, A_c is the cross-sectional area, and α and β are the short- and long-term prestress loss factors. It is assumed in Equations 9.1(a)–(d) that the transfer and service load bending moments, M_i and M_s respectively, are sagging moments. In sections of a member where either of these are hogging moments, the signs of M_i and M_s in Equations 9.1(a)–(d) must be reversed. If the maximum allowable compressive stresses in the concrete are f'_{max} and f_{max} at transfer and service loads, respectively, and the corresponding minimum allowable stresses are f'_{min} and f_{min}, respectively (note that if f_{min} is negative it would represent an allowable tensile stress), then Equations 9.1 may now be rewritten as inequalities:

$$\frac{\alpha P_i}{A_c} - \frac{\alpha P_i e}{Z_t} + \frac{M_i}{Z_t} \geqslant f'_{min} \tag{9.2a}$$

$$\frac{\alpha P_i}{A_c} + \frac{\alpha P_i e}{Z_b} - \frac{M_i}{Z_b} \leqslant f'_{max} \tag{9.2b}$$

$$\frac{\beta P_i}{A_c} - \frac{\beta P_i e}{Z_t} + \frac{M_s}{Z_t} \leqslant f_{max} \tag{9.2c}$$

$$\frac{\beta P_i}{A_c} + \frac{\beta P_i e}{Z_b} - \frac{M_s}{Z_b} \geqslant f_{min}. \tag{9.2d}$$

Inequalities 9.2(a)–(d) are shown graphically in Fig. 9.2. By combining inequalities 9.2(a) and 9.2(c), an expression for Z_t may be derived:

$$Z_t \geqslant \frac{(\alpha M_s - \beta M_i)}{(\alpha f_{max} - \beta f'_{min})}. \tag{9.3a}$$

Similarly, inequalities 9.2(b) and 9.2(d) may be combined to give an expression

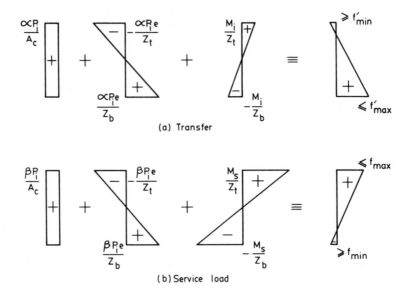

(a) Transfer

(b) Service load

Fig. 9.2 Design inequalities.

for Z_b:

$$Z_b \leqslant \frac{(\alpha M_s - \beta M_i)}{(\beta f'_{max} - \alpha f_{min})}. \tag{9.3b}$$

Note that the two expressions for the minimum values of Z_t and Z_b depend only on the *difference* between the maximum and minimum bending moments and allowable stresses, and not on their absolute values. These minima, however, take no account of practical values of prestress force and eccentricity. In practice, values for Z_t and Z_b larger than those given by inequalities 9.3(a) and 9.3(b) are usually chosen.

EXAMPLE 9.1   ■ ■

A post-tensioned prestressed concrete bridge deck is in the form of a solid slab (Fig. 9.3) and is simply supported over 20 m. It carries a service load of $10.3\,\text{kN/m}^2$. The allowable concrete stresses are given below. If the total short-

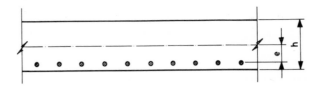

Fig. 9.3

and long-term losses are 10% and 20%, respectively, determine the minimum depth of slab required.

$$f'_{max} = 20.0 \, N/mm^2 \qquad f_{max} = 16.7 \, N/mm^2$$

$$f'_{min} = -1.0 \, N/mm^2 \qquad f_{min} = 0 \, N/mm^2.$$

$$M_i = 24h \times 20^2/8 = 1200h \, kN \, m/m,$$

where h is the overall depth of the slab in metres;

$$M_s = 1200h + (10.3 \times 20^2)/8$$
$$= (1200h + 515) \, kN \, m/m.$$

Thus, from inequalities 9.3(a) and (b),

$$Z_t \geqslant \frac{[0.9 \times (1200h + 515) - 0.8 \times 1200h]}{[(0.9 \times 16.7) - 0.8(-1)]} \times 10^6$$

$$= (7.58h + 29.28) \times 10^6 \, mm^3/m$$

$$Z_b \geqslant \frac{[0.9 \times (1200h + 515) - 0.8 \times 1200h]}{[(0.8 \times 20.0) - 0.9(0)]} \times 10^6$$

$$= (7.50h + 28.97) \times 10^6 \, mm^3/m.$$

For a rectangular section,

$$Z_t = Z_b = 10^3 \times (h^2/6) \times 10^6$$
$$= 0.167h^2 \times 10^9 \, mm^3/m.$$

Thus two equations can be formed for h:

$$0.167h^2 \times 10^9 = 7.58h + 29.28 \times 10^6$$
$$0.167h^2 \times 10^9 = 7.50h + 28.97 \times 10^6.$$

Solving these two equations gives values for h of 0.442 m and 0.440 m, respectively. Thus the minimum depth of the slab must be 442 mm. When initially sizing a section using inequalities 9.3(a) and 9.3(b), it is better to be on the generous side, in order to ensure that the ultimate limit state is satisfied. This will also ensure that the effects of misplaced tendons during construction will be minimized.

■ ■

9.3 Design of prestress force

The next stage in the design process is to find the prestress force, based on a maximum eccentricity determined from the section properties.

Rearranging inequalities 9.2(a)–(d) will yield inequalities for the required prestress force, for a given value of eccentricity. Thus the new inequalities are:

$$P_i \geqslant \frac{(Z_t f'_{min} - M_i)}{\alpha(Z_t/A_c - e)} \qquad (9.4a)$$

$$P_i \leqslant \frac{(Z_b f'_{max} + M_i)}{\alpha(Z_b/A_c + e)} \qquad (9.4b)$$

$$P_i \leqslant \frac{(Z_t f_{max} - M_s)}{\beta(Z_t/A_c - e)} \qquad (9.4c)$$

$$P_i \geqslant \frac{(Z_b f_{min} + M_s)}{\beta(Z_b/A_c + e)}. \qquad (9.4d)$$

There are thus two upper and two lower bounds on the value of the prestress force. Generally the minimum value of prestress force within these bounds is required, since the cost of the prestressing steel is a large proportion of the total cost of prestressed concrete structures.

EXAMPLE 9.2   ■ ■

For the bridge deck in Example 9.1, with a depth of 525 mm, if the maximum eccentricity of the tendons at midspan is 75 mm above the soffit, find the minimum value of the prestress force required.

$$Z_t = Z_b = 525^2 \times 10^3/6 = 45.94 \times 10^6 \, mm^3/m$$
$$A_c = 5.25 \times 10^5 \, mm^2/m$$
$$e = 525/2 - 75 = 188 \, mm$$
$$M_i = 1200 \times 0.525 = 630 \, kN \, m/m$$
$$M_s = 630 + 515 = 1145 \, kN \, m/m.$$

Inequalities 9.4(a)–(d) give the following values for P_i:

$$P_i \leqslant 7473.4 \, kN/m$$
$$P_i \leqslant 6246.3 \, kN/m$$
$$P_i \geqslant 4699.3 \, kN/m$$
$$P_i \geqslant 5195.0 \, kN/m.$$

The minimum value of P_i which lies within these limits is 5195.0 kN/m. Note that in inequalities 9.4(a) and 9.4(c) the denominator is negative for this example. Dividing both sides of an inequality by a negative number has the effect of changing the sense of the inequality.

If the prestress force is provided by evenly spaced tendons, each with an initial prestress force of 1378 kN, the spacing of the tendons is 265 mm. In this example, the prestress force can be varied easily by adjusting the tendon spacing. In the case of a beam, however, if the allowable range of prestress force given by inequalities 9.4(a)–(d) is small, it may be difficult to provide a practical arrangement of tendons which falls within this range. Unlike the case of reinforced concrete, where the over-provision of reinforcement only adds to the strength of a member, with prestressed concrete members too high a prestress force can lead to allowable stresses being exceeded at both transfer and service load.

The stresses at the two loading conditions are:
Transfer:

$$f_t = \frac{0.9 \times 5195.0 \times 10^3}{5.25 \times 10^5} - \frac{0.9 \times 5195.0 \times 10^3 \times 188}{45.94 \times 10^6} + \frac{630 \times 10^6}{45.94 \times 10^6}$$

$$= 8.91 - 19.13 + 13.71$$

$$= 3.49 \text{ N/mm}^2 (> f'_{min});$$

$$f_b = 8.91 + 19.13 - 13.71$$

$$= 14.33 \text{ N/mm}^2 (< f'_{max}).$$

Service:

$$f_t = (0.8/0.9) \times 8.91 - (0.8/0.9) \times 19.13 + (1145 \times 10^6)/(45.94 \times 10^6)$$

$$= 7.92 - 17.00 + 24.92$$

$$= 15.84 \text{ N/mm}^2 (< f_{max});$$

$$f_b = 7.92 + 17.00 - 24.92$$

$$= 0 \text{ N/mm}^2 (= f_{min}).$$

The most critical stress condition is that corresponding to Equation 9.1(d), the minimum stress condition under service load, while all the other stresses are within the prescribed limits. This is to be expected, since the prestressing force chosen was determined using Inequality 9.4(d).

■ ■

9.4 Magnel diagram

The four inequalities for the prestress force in Example 9.1 yielded a range of possible values for P_i. However, for a given value of e there may not be such a range, since the two inner of the four bounds to P_i could overlap. In this case, another value of e must be chosen and the limits to P_i found again, the process being repeated until a satisfactory combination of P_i and e is found. Clearly a more direct way of arriving at such a combination would be very useful.

To this end, Inequalities 9.4(a)–(d) may be written in the following form:

$$\frac{1}{P_i} \leqslant \frac{\alpha(Z_t/A_c - e)}{(Z_t f'_{min} - M_i)} \tag{9.5a}$$

$$\frac{1}{P_i} \geqslant \frac{\alpha(Z_b/A_c + e)}{(Z_b f'_{max} + M_i)} \tag{9.5b}$$

$$\frac{1}{P_i} \geq \frac{\beta(Z_t/A_c - e)}{(Z_t f_{max} - M_s)} \tag{9.5c}$$

$$\frac{1}{P_i} \leq \frac{\beta(Z_b/A_c + e)}{(Z_b f_{min} + M_s)}. \tag{9.5d}$$

As with the earlier inequalities, care must be taken with Inequalities 9.5(a) and 9.5(c). They are only valid if the denominators are positive. If either of the denominators are negative, then the original Inequality 9.2(a) or 9.2(c) has been divided by a negative quantity, and the inequality must be reversed.

The relationships between $1/P_i$ and e are linear and, if plotted graphically, they provide a very useful means of determining appropriate values of P_i and e. These diagrams were first introduced by the Belgian engineer, Magnel.

EXAMPLE 9.3   ■ ■

Construct a Magnel diagram for the bridge slab in Example 9.1 and find the minimum prestress force for a tendon eccentricity of 188 mm. What would be the effect on the minimum prestress force of: (i) reducing the eccentricity to 125 mm; (ii) increasing it to 250 mm?

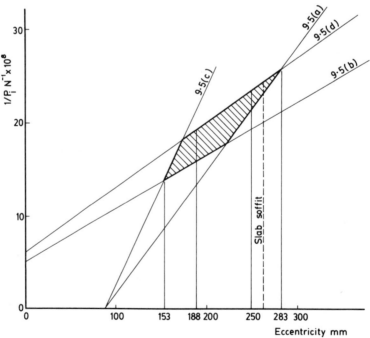

Fig. 9.4 Magnel diagram.

Inequalities 9.5(a)–(d) may be written as

$$10^8/P_i \geqslant 0.133e - 11.65$$
$$10^8/P_i \geqslant 0.058e + 5.08$$
$$10^8/P_i \leqslant 0.212e - 18.53$$
$$10^8/P_i \leqslant 0.070e + 6.11.$$

Note that the inequality signs in 9.5(a) and 9.5(c) have reversed since the denominators are negative. If the above inequalities are plotted with axes as $1/P_i$ and e, then each is a linear relationship defining a feasible region shown shaded in Fig. 9.4.

For any given eccentricity, it is easy to see which pair of Inequalities 9.4(a)–(d) will give the limits to P_i. For $e = 188$ mm, the range of allowable values for P_i is given by Inequalities 9.5(b) and (d), i.e.

$$P_i \leqslant 6246.3 \,\text{kN/m}$$
$$P_i \geqslant 5195.0 \,\text{kN/m}.$$

If the maximum eccentricity is reduced to 125 mm, it can be seen from the Magnel diagram that there is no feasible range for P_i. This means that it is impossible to satisfy the four Inequalities 9.2(a)–(d) with this value of e – at least one of the extreme fibre stresses must exceed the allowable value.

If the value of e is increased to 250 mm, the range of values for P_i is now given by Inequalities 9.4(a) and 9.4(d), i.e.

$$P_i \leqslant 4621.9 \,\text{kN/m}$$
$$P_i \geqslant 4240.7 \,\text{kN/m}.$$

■ ■

The value of e could be increased further, resulting in a range for P_i which would give smaller absolute values, but eventually the tendon position would reach the soffit of the slab. The feasible region extends from $e = 153$ to $e = 283$ mm. The two limits correspond to the overall maximum and minimum prestress force, respectively, in the section, and each is governed by the maximum and minimum stresses, respectively, under all load conditions. But in practice the limiting eccentricity is less than half the depth of the slab, due to the cover which must be provided to the prestressing steel. The value of $e = 188$ mm is the maximum practical eccentricity for this example, giving adequate cover from the soffit of the slab.

These variations of P_i with e show a general trend, namely that increasing e reduces P_i and *vice versa*. For minimum prestress force, maximum eccentricity should be provided at the point of maximum applied bending moment. This will ensure maximum ultimate strength, also.

The Magnel diagram is a very useful tool for understanding the relationship between prestress force and eccentricity. Even though much of the routine calculation work involved in prestressed concrete design is nowadays carried out by computer, it is essential for a designer to understand the way the variables in the design process affect one another.

9.5 Cable zone

Once the prestress force has been chosen based on the most critical section, it is possible to find the limits of the eccentricity e at sections elsewhere along the member. Thus an allowable *cable zone* is produced, within which the profile may take any shape. Here, the term 'cable' is used to denote the resultant of all the individual tendons. As long as the cable lies within the zone, the stresses at the different loading stages will not exceed the allowable values, even though some of the tendons might physically lie outside the cable zone.

Inequalities 9.5(a)–(d) may be rearranged to give

$$e \leqslant \frac{Z_t}{A_c} + \frac{1}{\alpha P_i}(M_i - Z_t f'_{min}) \tag{9.6a}$$

$$e \leqslant \frac{1}{\alpha P_i}(Z_b f'_{max} + M_i) - \frac{Z_b}{A_c} \tag{9.6b}$$

$$e \geqslant \frac{Z_t}{A_c} + \frac{1}{\beta P_i}(M_s - Z_t f_{max}) \tag{9.6c}$$

$$e \geqslant \frac{1}{\beta P_i}(Z_b f_{min} + M_s) - \frac{Z_b}{A_c}. \tag{9.6d}$$

EXAMPLE 9.4 ■ ■

In Example 9.3, if the prestress force is 5195.0 kN/m, determine the cable zone for the full length of the bridge deck, and a suitable cable profile.

The limits for the cable zone given by Inequalities 9.6(a)–(d) are:

$$e \leqslant 97.3 + 2.139 \times 10^{-7} M_i$$
$$e \leqslant 109.0 + 2.139 \times 10^{-7} M_i$$
$$e \geqslant -97.1 + 2.406 \times 10^{-7} M_s$$
$$e \geqslant -87.5 + 2.406 \times 10^{-7} M_s.$$

The values of M_i, M_s and e along the length of the slab are shown in Table 9.1 for one half of the slab, since all the values are symmetrical about the centreline. In this example, for all sections along the slab, Inequalities 9.6(a) and 9.6(d) will give the limits to the cable zone, which is shown in Fig. 9.5. This is usually the case if the minimum prestress force has been chosen, since these inequalities relate to the minimum stresses under all load conditions.

The width of the zone at the midspan section is 44 mm, which is sufficient to allow for any inaccuracies in locating the tendon ducts. However, for the chosen

Table 9.1 Cable zone for slab in Example 9.1.

Distance (m)	0	2.5	5.0	7.5	10.0
$M_i(kN\,m)$	0	275.6	472.5	590.6	630.0
$M_s(kN\,m)$	0	500.9	858.8	1073.4	1145.0
$e \geqslant (mm)$	−88	33	119	171	188
$e \leqslant (mm)$	97	156	198	224	232

prestress force of 5195.0 kN/m, one limit to the cable zone is $e = 188$ mm, which was fixed earlier as the maximum practical eccentricity for this structure. Thus, if the tendons are nominally fixed with an eccentricity of 188 mm, a small displacement upwards would bring the prestress force outside the cable zone. In order to overcome this, the spacing of the tendons is decreased slightly from 265 mm to 250 mm, giving an increased prestress force of 5512.0 kN/m. The limits to the cable zone at the midspan are now 172 mm and 224 mm, and so the nominal eccentricity of 188 mm lies within the cable zone with an acceptable tolerance of 16 mm.

If the shape of the chosen cable profile is parabolic, then if the eccentricity at midspan is 188 mm and at the support it is zero, giving a uniform stress at this point, the shape of the profile is given by

$$y = (4 \times 0.188/20^2)x(20 - x)$$

where y is a coordinate measured from the centroid of the section. The coordinates of the curve along the length of the deck can be found, and these are used to fix the tendon ducts in position during construction (Fig. 9.6). These coordinates can be shown to lie within the revised cable zone, based on $P_i = 5512.0$ kN/m.

■ ■

One important factor in choosing a cable profile for a post-tensioned member is the detail of the end-blocks. Manufacturers of the various prestressing systems usually specify the clearances required for their anchorages, and these will influence the eccentricity of the tendons at the end of the member. The design of end-blocks is discussed in Chapter 8.

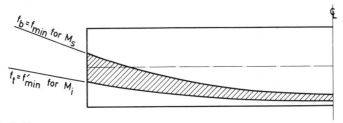

Fig. 9.5 Cable zone.

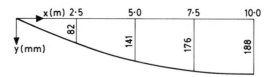

Fig. 9.6 Cable profile.

In the above example the magnitude of the prestressing force has been assumed to be constant; in real post-tensioned members the prestress force varies. An example illustrating both of these factors will be shown in Chapter 13.

9.6 Minimum prestress force

Since a significant portion of the cost of prestressed concrete members is in the prestressing steel, in any design the aim should be to reduce this to a minimum. Assuming that the steel is stressed to its limit, this is equivalent to keeping the prestress force to a minimum.

As shown in Section 9.3, to satisfy the basic inequalities concerning maximum and minimum concrete stresses at transfer and service load, there is usually a range within which the prestress force must lie. The minimum prestress force required for the service load in a simply supported beam is achieved when the eccentricity is a maximum. However, the eccentricity will be limited by consideration of the minimum concrete stress at transfer.

It is useful to examine the prestress force required in a given section, with a particular eccentricity but with varying transfer and service load bending moments. For the slab in Example 9.1. Inequalities 9.4(a)–(d) can be rearranged as

$$P_i \leqslant 11.056M_i + 508.0$$
$$P_i \leqslant 4.033M_i + 3705.6$$
$$P_i \geqslant 12.438M_s - 9542.7$$
$$P_i \geqslant 4.537M_s.$$

These inequalities are shown graphically in Fig. 9.7, which shows that for the maximum eccentricity of 188 mm the two limits to the prestress force are 5195.0 kN/m and 6246.3 kN/m, as determined previously. It also shows that the minimum prestress force is governed by Inequality 9.4(d) up to the value of 5479.6 kN/m. After this point, the prestress force is governed by Inequality 9.4(c), but more importantly, the rate of increase in service load bending moment M_s with prestress force is much reduced. The prestress force of 5479.6 kN/m may thus be regarded as an economic maximum force to provide. If a higher force is required for the given section, then probably it would be better to increase the section size.

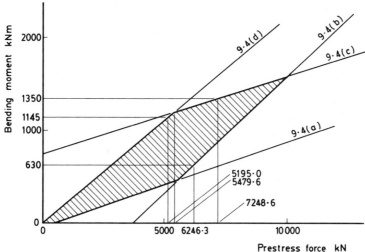

Fig. 9.7 Limits to prestress force.

It can also be seen from Fig. 9.7 that, for a given variation of bending moment $(M_s - M_i)$, there is a corresponding range of prestress force. If, in Example 9.1, M_i remains at $630\,\mathrm{kN\,m/m}$, but M_s is increased to $1350\,\mathrm{kN\,m/m}$, the limits to the prestress force are

$$P_i \leqslant 6246.3\,\mathrm{kN/m}$$
$$P_i \geqslant 7248.6\,\mathrm{kN/m}.$$

Clearly there is no feasible range for P_i, and the depth of the slab should be increased.

The ratio of M_s to M_i will also affect the minimum prestress force. For a given section, if this ratio is low, as is usually the case in long-span beams, the prestress force may be placed at a greater eccentricity, and hence may be smaller in value, than in the same section where the ratio of M_s to M_i is high.

For members which have a high ratio of M_s to M_i, one solution is to apply the prestress force in stages. This is carried out either by initially tensioning some, only, of the tendons to their full force, or by tensioning them all to a much lower initial force. In the latter case, the anchorage system must allow the tendons to be tensioned again to their full force at a later stage. Another alternative is to have the initial prestress force provided by pretensioned tendons, and the remaining prestress force supplied by post-tensioned tendons, tensioned at a later stage.

An example of where stage prestressing would be advantageous is in a building where a large clear span is required at ground floor level, and the columns from several upper floors are carried by a prestressed concrete beam at first floor level. If the beam were prestressed initially for the full service load, then only a small eccentricity at midspan could be tolerated to cater for the condition of minimum stress at transfer, leading to a large total prestress force. By tensioning the beam in stages, as each upper floor is added, the eccentricity at midspan could be increased, resulting in a smaller total prestress force.

9.7 Ultimate strength design

Once the details of the prestress force and cable profile have been determined, the ultimate limit state must be satisfied. If the ultimate strength is insufficient, then it will usually suffice to provide some extra untensioned reinforcement. The details of the analysis procedure for this case are given in Chapter 5.

The ultimate strength of a member at transfer is also important, but in practice this will usually be satisfactory if the serviceability limit state at transfer is satisfied.

EXAMPLE 9.5

For the bridge deck in Example 9.1, determine the ultimate moment of resistance of the section at midspan with $e = 188$ mm. Assume $f_{pu} = 1770$ N/mm^2, $f_{pi} = 1239$ N/mm^2, $f_{cu} = 40$ N/mm^2, $E_s = 195$ kN/mm^2 and that the total area of steel per metre $= 4449$ mm^2. Assume that the tendon ducts have been grouted after tensioning the tendons. Determine the amount of any extra reinforcement which may be required, with $f_y = 460$ N/mm^2.

The stress–strain curve for the particular grade of steel used is shown in Fig. 9.8, and the stress and strain distributions shown in Fig. 9.9.

The strain ε_{pe} in the prestressing steel at the ultimate limit state due to prestress only is given by

$$\varepsilon_{pe} = (0.8 \times 1239)/(195 \times 10^3) = 0.00508.$$

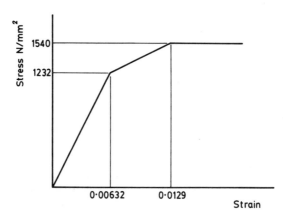

Fig. 9.8 Stress–strain curve for tendons.

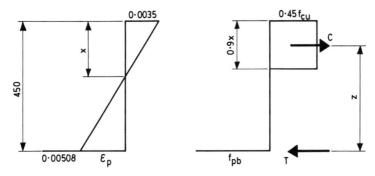

Fig. 9.9 Strain and stress distributions for slab in Example 9.5.

The total strain in the steel $\varepsilon_{pb} = \varepsilon_{pe} + \varepsilon_p$, and ε_p is determined from the strain diagram in Fig. 9.9:

$$0.0035/x = \varepsilon_p/(450 - x)$$

$$\therefore \varepsilon_p = (450 - x)(0.0035/x).$$

The stress in the steel is found from the stress–strain curve and the forces in the concrete and steel, C and T respectively, are then determined according to the principles shown in Section 5.4. Table 9.2 shows these forces for different values of x.

The neutral axis depth may thus be taken with sufficient accuracy to be 336 mm, showing that the steel has not yielded.

$$\text{Ultimate moment of resistance} = 5440(450 - 0.45 \times 336) \times 10^{-3}$$
$$= 1625.5 \, \text{kN m/m}$$

$$\text{Ultimate applied uniform load} = 1.4 \times 12.6 + 1.6 \times 10.3$$
$$= 34.1 \, \text{kN/m}^2.$$

Therefore, the maximum ultimate bending moment $= 34.1 \times 20^2/8$
$$= 1705 \, \text{kN m/m}.$$

Thus, extra untensioned reinforcement is required. The effective depth for this extra steel is $(525 - 50) = 475 \, \text{mm}$.

In order to obtain an estimate of the required amount of untensioned

Table 9.2 Neutral axis depth for slab in Example 9.5.

x (mm)	ε_p	ε_{pb}	f_{pb} (N/mm^2)	T (kN)	C (kN)
300	0.00175	0.00683	1256	5588	4860
330	0.00127	0.00635	1233	5486	5346
336	0.00119	0.00627	1223	5440	5443

reinforcement A_s, it may be assumed initially that both the prestressing steel and the untensioned reinforcement have not yielded, since the presence of any extra reinforcement in the section will lower the neutral axis. If the neutral axis depth is taken as approximately 370 mm, an equilibrium equation can be written to determine A_s.

$$0.45 \times 40 \times 10^3 \times 0.9 \times 370 = \{[(450 - 370)/370] \times 0.0035$$
$$+ 0.00508\} \times 195 \times 10^3 \times 4449$$
$$+ [(475 - 370)/370] \times 0.0035$$
$$\times 200 \times 10^3 A_s.$$
$$\therefore A_s = 4683 \text{ mm}^2/\text{m}.$$

This will be provided by T32 bars at 150 mm centres ($A_s = 5360 \text{ mm}^2/\text{m}$). It is now necessary to check that the ultimate moment of resistance of the section is greater than the applied bending moment. This is achieved, again, by using a trial-and-error procedure, as outlined in Table 9.3.

The strain in the untensioned reinforcement is given by

$$\varepsilon_{st} = (475 - x)(0.0035/x)$$

and the corresponding stress f_{st} is found from the stress–strain curve shown in Fig. 9.10.

Table 9.3

x (mm)	ε_p	ε_{pb}	ε_{st}	f_{pb} (N/mm²)	f_{st} (N/mm²)	C (kN)	T (kN)
370	0.00076	0.00584	0.00099	1139	198	6129	5994
373	0.00072	0.00580	0.00096	1131	192	6061	6043

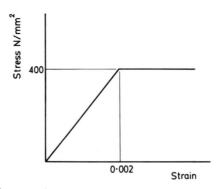

Fig. 9.10 Stress–strain curve for reinforcement.

The depth of the neutral axis is thus 373 mm, and the ultimate moment of resistance is given by

$$M_u = [4449 \times 1131(450 - 373)$$
$$+ 5360 \times 192(475 - 0.45 \times 373)] \times 10^{-6}$$
$$= 1735.8 \text{ kN m/m}.$$

This is greater than the ultimate applied bending moment, and therefore the section is satisfactory.

■ ■

9.8 Class 3 members

While the critical limit state for Class 1 and 2 members is generally that of serviceability, for Class 3 members the most critical is usually the ultimate limit state. Indeed, one way of viewing Class 3 members is as reinforced concrete members with sufficient prestress force applied to restrict the cracking under service load.

An approach to the design of Class 3 members, therefore, is to find the total area of steel required to give the desired ultimate moment of resistance, and then to proportion this total area between the prestressing steel and untensioned reinforcement. There are several criteria for determining the proportions for each type of steel.

One criterion is based on the hypothetical concrete tensile stresses given in BS8110, described in Chapter 5. The design process, once the ultimate limit state has been satisfied, is essentially the same as that for Class 1 and 2 members, and the details of the prestressing steel are determined using the methods outlined in the preceding sections.

An alternative method is to consider the member as having zero stress at the tensile face at the point of maximum applied bending moment, under some proportion of the total service load. The designer is free to choose this proportion, but a common criterion is to ensure zero tension under the permanent load only.

Instead of considering the state of stress within a member at the critical sections, emphasis may be placed on the deflection of the member, since this can be controlled by prestressing. If draped tendons are used, then the load balancing technique is very convenient in determining the prestress force required to give zero deflection under some proportion of the service load. This proportion is often taken as the permanent load, as with the previous method. For members with straight tendons, the load balancing method cannot be used, and in general the deflections cannot be made to be zero everywhere along the member. However, by ensuring zero deflection at the critical section, a reasonably level member will result.

EXAMPLE 9.6   ■ ■

The T-beam shown in Fig. 9.11 spans 15 m and carries an imposed load of 10 kN/m. Determine the amounts of prestressing steel and untensioned reinforcement required based on each of the following criteria: (a) BS8110 hypothetical concrete tensile stresses; (b) zero tension at midspan under permanent load comprising the dead load plus one-third of the imposed load; (c) zero deflection at midspan under permanent load. For all cases, assume that $f_{cu} = 40\,\text{N/mm}^2$, $f_{pu} = 1770\,\text{N/mm}^2$, $f_y = 460\,\text{N/mm}^2$, and that the long-term prestress losses are 20%.

The first step in each case is to determine the total area of steel required to give adequate ultimate strength.

$$\text{Beam self weight} = 24(0.7 \times 0.3 + 0.3^2)$$
$$= 7.2\,\text{kN/m};$$

$$\text{Ultimate uniform load} = 1.4 \times 7.2 + 1.6 \times 10$$
$$= 26.1\,\text{kN/m};$$

$$\text{Maximum ultimate bending moment} = 26.1 \times 15^2/8 = 734.1\,\text{kN m}.$$

Initially, it may be assumed that only the prestressing steel contributes to the ultimate moment of resistance. If the neutral axis is assumed to lie within the flange, two equations of equilibrium can be written:

$$0.45 \times 40 \times 700 \times 0.9x = 0.87 \times 1770 A_{ps}$$
$$0.45 \times 40 \times 700 \times 0.9x(525 - 0.45x) = 734.1 \times 10^6.$$

Solving these two equations gives $x = 140\,\text{mm}$ and $A_{ps} = 1031\,\text{mm}^2$.
The steel strain must now be checked:

$$\varepsilon_{pb} = \frac{0.8 \times 0.7 \times 1770}{195 \times 10^3} + \frac{(525 - 140)}{140} \times 0.0035$$

$$= 0.0147.$$

Since this is greater than ε_2 for the grade of steel used (see Fig. 9.8), the steel has yielded, as assumed.

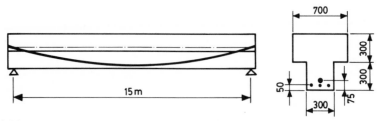

Fig. 9.11

(a) From Tables 5.3 and 5.4, for a limiting crack width of 0.2 mm, the hypothetical concrete tensile stress is $5.0 \times 0.9 = 4.5\,\text{N/mm}^2$.

$$M_s = 17.2 \times 15^2/8$$
$$= 483.8\,\text{kN m.}$$

Section properties:

$$Z_b = 22.00 \times 10^6\,\text{mm}^3$$
$$A_c = 3.00 \times 10^5\,\text{mm}^2$$
$$e = 285\,\text{mm.}$$

For the bottom fibre stress at midspan under service load.

$$\frac{0.8P_i \times 10^3}{3.00 \times 10^5} + \frac{0.8P_i \times 10^3 \times 285}{22.00 \times 10^6} - \frac{483.8 \times 10^6}{22.00 \times 10^6} = -4.5;$$

$$\therefore P_i = 1342.3\,\text{kN.}$$

If the initial stress in the tendon is $0.7f_{pu}$, the area of prestressing steel required is given by

$$A_{ps} = \frac{1342.3 \times 10^3}{0.7 \times 1770} = 1083\,\text{mm}^2.$$

This is greater than the area of prestressing steel required for the ultimate limit state, and so no additional untensioned reinforcement is required.

(b) For zero tension at the beam soffit at midspan under permanent load, the maximum bending moment is given by

$$M_{per} = (7.2 + 10/3) \times 15^2/8 = 296.3\,\text{kN m.}$$

The required prestress force is then given by

$$\frac{0.8P_i \times 10^3}{3.00 \times 10^5} + \frac{0.8P_i \times 10^3 \times 285}{22.00 \times 10^6} - \frac{296.3 \times 10^6}{22.00 \times 10^6} = 0;$$

$$\therefore P_i = 1033.6\,\text{kN}$$

$$\therefore A_{ps} = \frac{1033.6 \times 10^3}{0.7 \times 1770} = 834\,\text{mm}^2.$$

Since this area of steel is less than the amount of prestressing steel required for the ultimate limit state, additional untensioned reinforcement is required.
Again, two equilibrium equations can be formed:

$$0.45 \times 40 \times 700 \times 0.9x = 0.87 \times 1770 \times 834 + 0.87 \times 460A_s$$
$$0.87 \times 1770 \times 834(525 - 0.45x) + 0.87 \times 460A_s(550 - 0.45x) = 734.1 \times 10^6.$$

Solving these equations gives $x = 138\,\text{mm}$ and $A_s = 701\,\text{mm}^2$. Checks on the steel strains show that both types of steel have yielded, as assumed.

(c) For zero deflection under the permanent load, the required prestressing force may be found conveniently using the load balancing technique. If the tendon has zero eccentricity at the supports, then the total drape is 285 mm.

Load to be balanced $= 7.2 + 10/3 = 10.5\,\text{kN/m}$.

Thus

$$0.8 P_i = (10.5 \times 15^2)/(8 \times 0.285);$$

$$\therefore P_i = 1295.2\,\text{kN};$$

$$\therefore A_{ps} = \frac{1295.2 \times 10^3}{0.7 \times 1770} = 1045\,\text{mm}^2.$$

In this case, no extra reinforcement is required.   ■ ■

Whichever method is used, the design should be completed by checking the concrete compressive stresses at transfer and under service load against the allowable stresses given in Chapter 3. The serviceability limit state of cracking should be checked either by determining the crack widths directly at transfer and under service load, using Equation 5.3, or by finding the stress in the steel adjacent to the tension face (using a cracked-section analysis) and comparing it with the allowable values given in Section 5.11.

The three alternative designs in Example 9.6 illustrate the fact that, with Class 3 members, the designer has great freedom to choose the prestress force to suit any required criterion, but careful attention must always be paid to the serviceability limit state. In Example 9.6, the imposed load was high relative to the dead load and the permanent load thus included a portion of the imposed load. If the imposed load is low in relation to the dead load, then the permanent load is often taken as the dead load only.

It is interesting to note that if the beam in Example 9.6 had been designed as a Class 1 or a Class 2 member, the areas of prestressing steel required, based on the bottom fibre concrete stresses under service load, would be 1362 mm² and 1220 mm² respectively, and in neither case is any additional untensioned reinforcement required.

For a given structure, the choice of which class of member to use depends on the function of the structure and the nature of the loading. Where it is important to have a crack-free structure, such as in a liquid-retaining structure, or where the environment is particularly aggressive, Class 1 members should be used. For structures with a high ratio of imposed to dead load, however, specifying Class 1 members may be too onerous, since larger amounts of prestressing steel would be required than if much of the load were permanent. Excessive camber may also be present in Class 1 members under their own weight.

Class 2 members will give the most economical structure for most purposes. Further savings in prestressing steel could be made, where limited cracking is acceptable, by using Class 3 members. They are particularly suitable where the maximum service load is only rarely reached, since any cracks which open up under this load would close again once the load has been reduced, and the structure would remain uncracked for most of its service life. Class 3 members are particularly useful for structures subjected to impact loading. They deflect more, and thus absorb more energy, than Class 1 and 2 members, and exhibit better elastic recovery after impact than reinforced concrete members.

A common concern is that partially prestressed members are less durable than fully prestressed members, since they are generally cracked under service load, but many partially prestressed concrete structures have been in service for more than 25 years and have behaved satisfactorily.

9.9 Choice of section

In the bridge deck slab shown in the previous examples, the simplest shape of cross-section, namely rectangular, was used, chosen primarily to illustrate the basic principles of design. Where there is freedom to choose a more economical section, the designer must decide which shape of section to use for a particular situation.

The solid rectangular section is one of the least economical sections, since the mid-depth regions are usually not highly stressed and the material is not being used to its full extent. One way of overcoming this deficiency is to provide voids in the central region of the section, which allow a similar structural efficiency with less weight. A typical hollow-core slab is shown in Fig. 9.12(a). As with steel sections, an I-section is a very efficient shape, Fig. 9.12(b), providing maximum area of concrete at the furthest distance from the neutral axis. An alternative

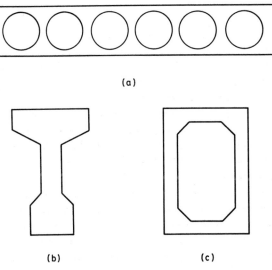

(a)

(b)   (c)

Fig. 9.12 Sections.

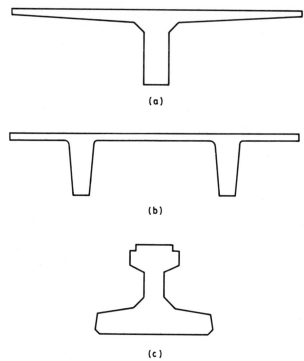

Fig. 9.13 T-sections.

section, one with similar efficiency for bending, but with far greater torsional stiffness, is the box section, shown in Fig. 9.12(c).

The T-section shown in Fig. 9.12(b) is suitable for long-span beams, generally in bridges. For buildings, the T-section shown in Fig. 9.13(a) is often used. This has a large compression flange for the service load, but it is necessary to ensure that the compressive stresses in the rib at transfer are not excessive. If the rib is slender, the possibility of buckling at transfer must also be considered. Single T-sections are not very stable during construction, and a common solution is to combine two of them into stable double T-sections, shown in Fig. 9.13(b).

Another shape which is often used in composite construction is the inverted T-section shown in Fig. 9.13(c). The large compression flange at the soffit of the beam can accommodate a large bending moment due to the self weight of the beam and the weight of the *in situ* slab, and, for the service load, the compression flange is supplied by the slab which acts compositely with the inverted T-section.

9.10 Flow charts for design

The methods of design of Class 1, 2 and 3 members outlined in the preceding sections of this chapter may be combined with the design elements considered in the previous chapters to give an overall view of the design process. This is most conveniently summarized in the form of flow charts, and Figs. 9.14 and 9.15 show these for Class 1 and 2, and for Class 3, members respectively.

Many steps in the design procedures are common for each class of member, and the main differences are in the determination of the details of the prestressing steel. For Class 1 and 2 members, emphasis is usually placed on the

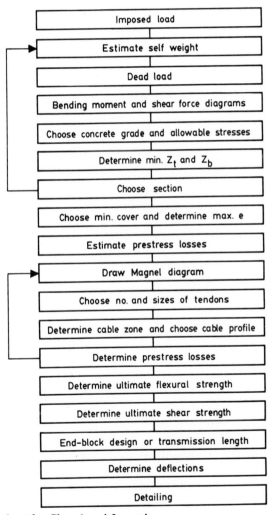

Fig. 9.14 Flow chart for Class 1 and 2 members.

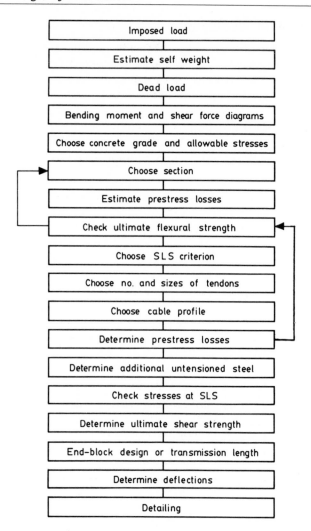

Fig. 9.15 Flow chart for Class 3 members.

stresses at the serviceability limit state, with checks for ultimate strength made afterwards. For Class 3 members, the ultimate strength capacity should be ensured first, and then conditions at the serviceability limit state checked later.

The steps shown in Figs. 9.14 and 9.15 are intended only as a guide, and with experience many of them may be combined or bypassed completely.

9.11 Detailing

There are some practical details concerning the layout of tendons that may affect the design, and it is important to be aware of these when deciding on the number and shape of tendons to be provided.

A certain minimum percentage of prestressing steel is required in a member to ensure that, when the concrete cracks, the additional force transferred to the steel would not cause immediate yield or rupture. This requirement may be considered to be satisfied if the ultimate moment of resistance is given by

$$M_u \geqslant (f_{pt} + 0.6 f_{cu}^{1/2}) I/y$$

where f_{pt} is the prestress at the tensile face of the member (a distance y from the centroid) after all losses have occurred.

A factor affecting the choice of numbers and sizes of individual tendons is the clear space which must be provided between tendons, in order to ensure proper placement and compaction of the surrounding concrete. The clear distance between post-tensioning ducts should not be less than the greatest of: (a) the maximum aggregate size plus 5 mm; (b) in the vertical direction, the vertical internal duct dimension; (c) in the horizontal direction, the horizontal internal duct dimension.

If two or more rows of ducts are used, the gaps between the ducts should be aligned to allow a vibrator to be inserted more easily, and sufficient horizontal clearance between the ducts should also be allowed for this.

For pretensioning tendons, the clear space requirements are similar to those for reinforcing steel, namely either the maximum aggregate size plus 5 mm, or the tendon diameter, whichever is the greater.

If a curved post-tensioned duct is placed near the surface of a concrete member, bursting of the concrete may occur in a direction perpendicular to the plane of curvature of the duct. In order to prevent this, the minimum cover to the duct should not be less than the value given in Table 9.4 for the given duct size and radius of curvature. In order to prevent the crushing of the concrete between curved post-tensioning ducts which are in the same plane of curvature, the clear spacing between the ducts in this plane should not be less than the values given in Table 9.5.

The minimum cover to ducts and tendons is usually determined from durability and fire resistance considerations, and is described in Chapter 3. Most manufacturers of prestressing systems also specify the minimum cover to be used.

Most prestressed concrete members will contain untensioned reinforcement fabricated into a cage. This serves several purposes: (a) to facilitate the placing of post-tensioning ducts; (b) to enhance the ultimate flexural and shear strength of the member; (c) to resist any tensile stresses which may be set up by restraint of shrinkage of the member by the formwork before the member is prestressed;

Table 9.4 Minimum cover (mm) to curved ducts.

Radius of curvature of duct (m)	Duct internal diameter (mm) 19	30	40	50	60	70	80	90	100	110	120	130	140	150	160	170
Tendon force $(0.8 f_{pu} \cdot A_{ps})$ (kN)	296	387	960	1337	1920	2640	3360	4320	5183	6019	7200	8640	9424	10338	11248	13200
2	50	55	155	220	320	445										
4		50	70	100	145	205	265	350	420	310						
6			50	65	90	125	165	220	265	220	375	460				
8				55	75	95	115	150	185	165	270	330	360	395		
10				50	65	85	100	120	140	145	205	250	275	300	330	
12					60	75	90	110	125	130	165	200	215	240	260	315
14					55	70	85	100	115	125	150	170	185	200	215	260
16					55	65	80	95	110	115	140	160	175	190	205	225
18					50	65	75	90	105	110	135	150	165	180	190	215
20						60	70	85	100	105	125	145	155	170	180	205
22						55	70	80	95	100	120	140	150	160	175	195
24						55	65	80	90	100	115	130	145	155	165	185
26						50	65	75	85	95	110	125	135	150	160	180
28							60	75	85	90	105	120	130	145	155	170
30							60	70	80	90	105	120	130	140	150	165
32							55	70	80	90	100	115	125	135	145	160
34							55	65	75	85	100	110	120	130	140	155
36							55	65	75	85	95	105	115	125	140	150
38							50	60	70	80	90	100	115	125	135	150
40	50	50	50	50	50	50	50	60	70	80	90	100	110	120	130	145

Radii not normally used (upper-right region of the table).

Table 9.5 Minimum distance (mm) between ducts in plane of curvature.

Radius of curvature of duct (m)	Duct internal diameter (mm)															
	19	30	40	50	60	70	80	90	100	110	120	130	140	150	160	170
	Tendon force $(0.8 f_{pu} A_{ps})(kN)$															
	296	387	960	1337	1920	2640	3360	4320	5183	6019	7200	8640	9424	10 336	11 248	13 200
2	110	140	350	485	700	960										
4	55	70	175	245	350	480	610	785	940					Radii not normally used		
6	38	60	120	165	235	320	410	525	630	730	870	1045				
8			90	125	175	240	305	395	470	545	655	785	855	940		
10			80	100	140	195	245	315	375	440	525	630	685	750	815	
12						160	205	265	315	365	435	525	570	625	680	800
14						140	175	225	270	315	375	450	490	535	585	785
16							160	195	235	275	330	395	430	470	510	600
18								180	210	245	290	350	380	420	455	535
20									200	220	265	315	345	375	410	480
22											240	285	310	340	370	435
24												265	285	315	340	400
26												260	280	300	320	370
28																345
30																340
32																
34																
36																
38																
40	38	60	80	100	120	140	160	180	200	220	240	260	280	300	320	340

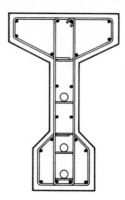

Fig. 9.16

(d) to enable the member to withstand any sudden load applied to it (the reinforcement should preferably be mild steel).

The detailing of the untensioned reinforcement is covered in the relevant sections of BS8110. An example of a reinforcement cage in a beam is shown in Fig. 9.16.

Chapter 10

COMPOSITE CONSTRUCTION

10.1 Introduction

Many applications of prestressed concrete involve the combination of precast prestressed concrete beams and *in situ* reinforced concrete slabs. Some examples of such composite construction are shown in Fig. 10.1. An *in situ* infill between precast beams is shown in Fig. 10.1(a) while an *in situ* topping is shown in Fig. 10.1(b). The difference in behaviour between these two types of slab will be discussed later. The beams are designed to act alone under their own weight plus the weight of the wet concrete of the slab. Once the concrete in the slab has hardened, provided that there is adequate horizontal shear connection between the slab and beam, they behave as a composite section under service load. The beams acts as permanent formwork for the slab, which provides the compression flange of the composite section. The section size of the beam can thus be kept to a minimum, since a compression flange is only required at the soffit at transfer. This leads to the use of inverted T-, or 'top-hat', sections.

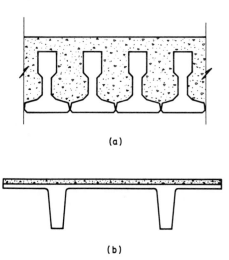

(a)

(b)

Fig. 10.1 Examples of composite construction.

10.2 Serviceability limit state

The stress distributions in the various regions of the composite member are shown in Figs 10.2(a)–(d). The stress distribution in Fig. 10.2(a) is due to the self weight of the beam and is similar to that described in Chapter 5, with the maximum compressive stress at the lower extreme fibre. Once the slab is in place, the stress distribution in the beam is modified to that shown in Fig. 10.2(b), where the bending moment at the section is that due to the combined self weight of the beam and slab, M_d.

Once the concrete in the slab has hardened and the imposed load acts on the composite section, the additional stress distribution is shown in Fig. 10.2(c). This is determined by ordinary bending theory, but using the composite section properties. The final stress distribution is shown in Fig. 10.2(d) and is a superposition of those shown in Fig. 10.2(b) and Fig. 10.2(c). The important feature to note is that there is a discontinuity in the final stress distribution under service load at the junction between the beam and slab. The beam has an initial stress distribution before it behaves as part of the composite section, whereas the slab only has stresses induced in it due to the composite action.

EXAMPLE 10.1   ■ ■

The floor slab shown in Fig. 10.3 comprises precast pretensioned beams and an *in situ* concrete slab. If the span of the beams is 5 m and the superimposed load is

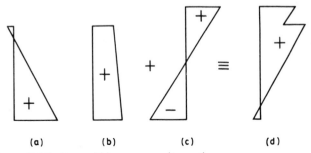

(a)   (b)   (c)   (d)

Fig. 10.2 Stress distribution within a composite section.

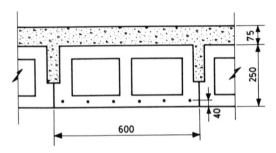

Fig. 10.3

$5 \, \text{kN/m}^2$ (including finishes), determine the stress distributions at the various load stages. Assume all long-term losses have occurred before the beams are erected and that the net force in each wire is $19.4 \, \text{kN}$.

Section properties of the beams:

$$A_c = 1.13 \times 10^5 \, \text{mm}^2$$
$$I = 7.5 \times 10^8 \, \text{mm}^4$$
$$Z_t = Z_b = 6.0 \times 10^6 \, \text{mm}^3.$$

Eccentricity of the wires $= 125 - 40 = 85 \, \text{mm}$.

(i) Self weight of the beams $= 0.113 \times 24$
$$= 2.7 \, \text{kN/m}.$$
$$M_i = (2.7 \times 5.0^2)/8$$
$$= 8.4 \, \text{kN m}.$$

Total prestress force after all losses have occurred is given by

$$\beta P_i = 6 \times 19.4$$
$$= 116.4 \, \text{kN}.$$

The stress distribution in the beams is thus given by

$$f_t = \frac{116.4 \times 10^3}{1.13 \times 10^5} - \frac{116.4 \times 85 \times 10^3}{6.0 \times 10^6} + \frac{8.4 \times 10^6}{6.0 \times 10^6}$$
$$= 1.03 - 1.65 + 1.40$$
$$= 0.78 \, \text{N/mm}^2;$$
$$f_b = 1.03 + 1.65 - 1.40$$
$$= 1.28 \, \text{N/mm}^2.$$

(ii) The weight of the slab is supported by the beams acting alone, so that

$$M_d = 8.4 + 0.075 \times 0.6 \times 24 \times 5.0^2/8$$
$$= 11.8 \, \text{kN m}.$$

The stress distribution within the beams is now given by

$$f_t = 1.03 - 1.65 + \frac{11.8 \times 10^6}{6.0 \times 10^6}$$
$$= -0.62 + 1.97$$
$$= 1.35 \, \text{N/mm}^2;$$
$$f_b = 1.03 + 1.65 - 1.97$$
$$= 0.71 \, \text{N/mm}^2.$$

(iii) The imposed load of $5 \, \text{kN/m}^2$ is supported by the composite section and the section properties of this are now required. To find the neutral axis of the composite section, taking moments about the soffit of the beams gives

$$(1.13 \times 10^5 + 75 \times 600)\bar{y} = (1.13 \times 10^5 \times 125 + 75 \times 600 \times 288)$$

$$\therefore \bar{y} = 171\,\text{mm}.$$

$$I_{comp} = 7.5 \times 10^8 + 1.13 \times 10^5\,(171 - 125)^2$$
$$+ (75^3 \times 600)/12 + (75 \times 600)/(288 - 171)^2$$
$$= 1.63 \times 10^9\,\text{mm}^4.$$

The imposed load bending moment $= 0.6 \times 5.0 \times 5.0^2/8$

$$= 9.4\,\text{kN m}.$$

The stress distribution within the composite section under this extra bending moment is given by

$$f_{t,slab} = \frac{9.4 \times 10^6}{1.63 \times 10^9} \times (325 - 171) = 0.89\,\text{N/mm}^2$$

$$f_{t,beam} = \frac{9.4 \times 10^6}{1.63 \times 10^9} \times (250 - 171) = 0.46\,\text{N/mm}^2$$

$$f_{b,beam} = \frac{-9.4 \times 10^6}{1.63 \times 10^9} \times 171 = -0.99\,\text{N/mm}^2.$$

The total stress distributions under the three load cases are shown in Fig. 10.4.

■ ■

The maximum compressive stress occurs at the upper fibres of the beams, but this has been significantly reduced from the level of stress had the beam carried the total service load alone. This explains the advantage of inverted T-sections in composite construction, where only a small compression flange is required

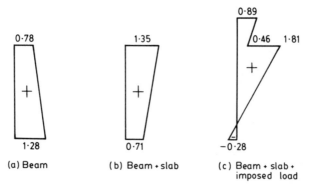

(a) Beam   (b) Beam + slab   (c) Beam + slab + imposed load

Fig. 10.4 Stress distribution for beam in Example 10.1.

for bending moments M_i and M_d, the compression flange for bending moment M_s being provided by the slab.

The allowable compressive stresses for the upper fibres in the beams acting compositely may be increased by 50% over those given in Table 3.3, since there is a confining effect from the *in situ* slab. This increase should only be applied if the flange of the beam is completely encased by *in situ* concrete, as in Fig. 10.1(a); otherwise the increase should be limited to 25%. The overall failure of the composite member must be by excessive elongation of the steel for this extra stress to be relied upon; that is, the section must be under-reinforced (see Chapter 5). The complete analysis of the floor beams should also, of course, include consideration of the stresses at transfer.

The *in situ* slab in Example 10.1 lies above the composite section neutral axis, and therefore the slab is in compression over its full depth under the service load. However, for composite sections as shown in Fig. 10.5, the *in situ* portion of the section extends well below the neutral axis, so that the lower region is in tension. If the tensile strength of this concrete is exceeded, the composite section properties must be determined on the basis of the *in situ* section having cracked below the neutral axis.

EXAMPLE 10.2 ■■

The composite bridge deck shown in Fig. 10.5 has a span of 15 m and is composed of inverted T-beams at 500 mm centres, with an overall depth of 845 mm. If the total prestressing force in each beam is 1140 kN after all losses have occurred, determine the stress distribution under an imposed load of 12 kN/m.

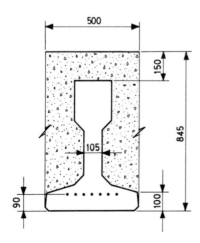

Fig. 10.5

Section properties of the inverted T-section:

$$I_b = 7.78 \times 10^9 \text{ mm}^4$$
$$A_c = 1.47 \times 10^5 \text{ mm}^2$$
$$Z_t = 19.20 \times 10^6 \text{ mm}^3$$
$$Z_b = 26.91 \times 10^6 \text{ mm}^3$$
$$w = 3.5 \text{ kN/m}$$
$$\bar{y} = 289 \text{ mm from soffit}$$
$$e = (289 - 90) = 199 \text{ mm}.$$

Weight of *in situ* concrete $= (0.5 \times 0.845 - 0.147) \times 24$
$$= 6.6 \text{ kN/m}.$$

$$M_i = 3.5 \times 15.0^2/8$$
$$= 98.4 \text{ kN m}.$$

The stress distribution in the beam under its own weight is given by

$$f_t = \frac{1140 \times 10^3}{1.47 \times 10^5} - \frac{1140 \times 10^3 \times 199}{19.20 \times 10^6} + \frac{98.4 \times 10^6}{19.20 \times 10^6}$$
$$= 7.76 - 11.82 + 5.13$$
$$= 1.07 \text{ N/mm}^2;$$
$$f_b = 7.76 + (11.82 - 5.13) \times (19.20/26.91)$$
$$= 12.53 \text{ N/mm}^2.$$

The extra bending moment due to the weight of the slab

$$= 6.6 \times 15.0^2/8$$
$$= 185.6 \text{ kN m}.$$

The stress distribution under the bending moment M_d is thus given by

$$f_t = 1.07 + \frac{185.6 \times 10^6}{19.20 \times 10^6}$$
$$= 10.74 \text{ N/mm}^2;$$
$$f_b = 12.53 - \frac{185.6 \times 10^6}{26.91 \times 10^6}$$
$$= 5.63 \text{ N/mm}^2.$$

The imposed load of 12 kN/m is supported by the composite section, which is rectangular with dimensions 845×500 mm.
Thus

$$\bar{y} = 423 \text{ mm, and}$$
$$I_{comp} = (845^3/12) \times 500$$
$$= 2.51 \times 10^{10} \text{ mm}^4.$$

The imposed load bending moment $= 12 \times 15^2/8$

$$= 337.5 \, \text{kN m.}$$

The stress distribution in the composite section under the imposed load is given by

$$f_{t,\text{slab}} = \frac{337.5 \times 10^6 \times 423}{2.51 \times 10^{10}}$$

$$= 5.69 \, \text{N/mm}^2;$$

$$f_{t,\text{beam}} = \frac{337.5 \times 10^6 \times 272}{2.51 \times 10^{10}}$$

$$= 3.67 \, \text{N/mm}^2;$$

$$f_{b,\text{beam}} = -\frac{337.5 \times 10^6 \times 423}{2.51 \times 10^{10}}$$

$$= -5.69 \, \text{N/mm}^2.$$

The stress distributions in the composite section under the various load cases are shown in Fig. 10.6.

The maximum compressive stress in the slab is much lower than in the beam, and for this reason, in many composite structures, a lower grade of concrete is used for the *in situ* portion. The modulus of elasticity for this concrete will be lower than that for the beam, and this effect can be taken into account in finding the composite section properties by using an approximate modular ratio of 0.8.

The tensile stress at the interface between the slab and the lower flange of the beam is given by

$$f_{b,\text{slab}} = -\frac{337.5 \times 10^6}{2.51 \times 10^{10}} (423 - 100)$$

$$= -4.34 \, \text{N/mm}^2.$$

■ ■

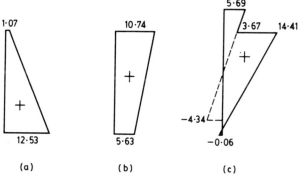

(a)   (b)   (c)

Fig. 10.6 Stress distribution for beam in Example 10.2 (N/mm²).

Table 10.1 Allowable tensile stresses in *in situ* concrete.

Grade of in situ concrete	Maximum tensile stress (N/mm^2)
25	3.2
30	3.6
40	4.4
50	5.0

The allowable compressive stresses in the slab and beam concrete are found from Table 3.3. If the slab and beam in Example 10.2 are of grade 30 and 50 concrete respectively, the allowable compressive stresses are 10.00 and 16.67 N/mm^2 respectively. The allowable tensile stress in the beam, from Table 3.3, is 3.18 N/mm^2. The allowable tensile stress in the slab may be taken from Table 10.1, for prestressed concrete beams with *in situ* in fills as in Fig. 10.5. For this case, the allowable stress is 3.6 N/mm^2.

The maximum stress in the *in situ* concrete thus exceeds the allowable stress. However, it is stated in BS8110 that this may occur, by up to 50%, provided the design tensile stress in the beam is reduced by a similar numerical amount. Thus, in Example 10.2, the allowable tensile stress in the slab may be increased by 1.8 N/mm^2 to 5.4 N/mm^2, provided that the allowable tensile stress at the lower flange of the beam is reduced to $(3.18 - 1.8) = 1.38$ N/mm^2. The actual stress here is compressive, so that the overall stress distribution is satisfactory. The justification for the increased allowable tension in the slab comes from experiments which show that the uncracked prestressed concrete beam inhibits the formation of cracks in the adjacent *in situ* concrete, and that this effect is enhanced with increase in level of prestress in the beam.

Had the allowable tensile strength of the slab been exceeded, even allowing for the increase described above, the composite section must be treated as cracked, and the section properties determined accordingly, using the methods described in Chapter 5.

For prestressed concrete beams with *in situ* toppings as in Fig. 10.7(a), the rules for detailing reinforcement in slabs given in BS8110 should be followed, in order to limit the cracking.

10.3 Ultimate strength

The basic principles for the analysis of prestressed concrete sections at the ultimate limit state of flexural strength described in Chapter 5 are also applicable to composite sections. For the section shown in Fig. 10.7(a), it may be assumed initially that, at the ultimate limit state, the neutral axis lies within the slab and the section may then be treated effectively as a rectangular beam. The

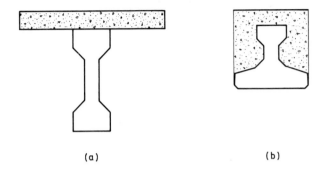

(a)   (b)

Fig. 10.7

position of the neutral axis should later be checked to see whether it does, indeed, fall within the slab.

For the section shown in Fig. 10.7(b), the position of the neutral axis may be determined on the assumption that the section is rectangular, but the different strengths of the concrete in the slab and beam regions of the compression zone should be taken into account.

EXAMPLE 10.3

Determine the ultimate moment of resistance of the section in Example 10.2, if $f_{pu} = 1770\,\text{N/mm}^2$, $f_{pe} = 990\,\text{N/mm}^2$, $A_{ps} = 1152\,\text{mm}^2$ and $E_s = 195\,\text{kN/mm}^2$.

The strain and stress distributions in the composite section at the ultimate limit state are shown in Fig. 10.8.

$$\varepsilon_{pe} = 990/(195 \times 10^3) = 0.00508.$$

Assuming   that   the   section   is   under-reinforced,   that   is   that

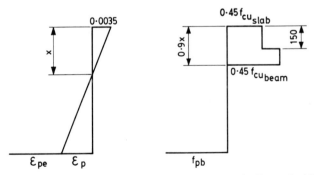

Fig. 10.8 Ultimate strain and stress distributions for beam in Example 10.2.

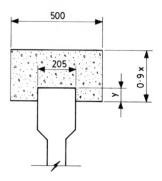

Fig. 10.9

$(\varepsilon_{pe} + \varepsilon_p) > \varepsilon_2 = 0.0129$, determined from the stress–strain diagram in Fig. 3.4, the stress in the prestressing steel is $0.87 f_{pu}$.

For equilibrium of forces within the section:

$$0.87 f_{pu} A_{ps} = (0.45 f_{cu} A_c)_{slab} + (0.45 f_{cu} A_c)_{beam}. \tag{10.1}$$

If the neutral axis lies within the upper flange of the beam, the values of A_c for the slab and beam can be determined from Fig. 10.9.

$$A_{c,slab} = 500\,(150 + y) - 205y$$
$$= 75000 + 295y;$$
$$A_{c,beam} = 205y.$$

Substituting these values into Equation 10.1 gives

$$0.87 \times 1770 \times 1152 = 0.45 \times 30\,(75000 + 295y) + 0.45 \times 50 \times 205y;$$
$$\therefore y = 89\,\text{mm}.$$

In order to check that the prestressing steel has yielded, the neutral axis must be found. This is given by

$$x = (150 + 89)/0.9$$
$$= 265\,\text{mm}.$$

From the strain diagram shown in Fig. 10.8,

$$\varepsilon_p = [(755 - 265)/265] \times 0.0035$$
$$= 0.00647.$$

Thus,

$$\varepsilon_{pb} = \varepsilon_{pe} + \varepsilon_p$$
$$= 0.00508 + 0.00647$$
$$= 0.01155.$$

This is less than the yield strain of 0.0129 and the original assumption that the steel has yielded is therefore no longer valid. The stress in the steel for a given strain ε_{pb} is now given by the stress–strain curve in Fig. 3.4. It can be shown that the neutral axis depth is 258 mm, with $y = 82$ mm.

The moment of resistance of the section can be found by taking moments about the prestressing steel. For the two areas of concrete in compression:

$$A_{c,slab} = 75000 + 295 \times 82$$
$$= 0.99 \times 10^5 \, mm^2;$$
$$A_{c,beam} = 205 \times 82$$
$$= 1.68 \times 10^4 \, mm^2.$$

For the areas of slab and beam in compression, the centroids are at a distance of 103 mm and 191 mm, respectively, from the top of the section. Thus the moment of resistance of the composite section is given by

$$M_u = [0.99 \times 10^5 \times 0.45 \times 30 \, (755 - 103)$$
$$+ 1.68 \times 10^4 \times 0.45 \times 50 \, (755 - 191)] \times 10^{-6}$$
$$= 1084.6 \, kN \, m.$$

■ ■

If necessary, the effect of additional untensioned reinforcement can be taken into account, as described in Chapter 5. The ultimate strength of the precast beam supporting its own weight plus that of the slab should also be checked.

10.4 Horizontal shear

The composite behaviour of the precast beam and *in situ* slab is only effective if the horizontal shear stresses at the interface between the two regions can be resisted. For shallow members, such as that shown in Fig. 10.3, there is usually no mechanical key between the two types of concrete, and reliance is made on the friction developed between the contact surfaces. For deeper sections, mechanical shear connectors in the form of links projecting from the beam are used, which provide a much better shear connection.

The determination of the horizontal shear resistance is based on the ultimate limit state, and if this condition is satisfied it may be assumed that satisfactory horizontal shear resistance is provided at the serviceability limit state.

A simply supported composite section carrying a uniformly distributed load is shown in Fig. 10.10(a) and the free-body diagram for half the length of the *in situ* slab is shown in Fig. 10.10(b). At the simply supported end there must be zero force in the slab, while the maximum force occurs at the midspan. The distribution of shear forces on the underside of the slab is also shown in Fig. 10.10(b), being zero at midspan and reaching a maximum at the support. This behaviour is similar to that in an elastic beam, where the vertical and

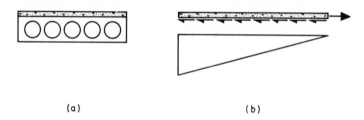

(a)   (b)

Fig. 10.10 Horizontal shear.

horizontal shear stresses increase towards the support for a uniformly distributed load.

The actual distribution of the ultimate shear force on the underside of the slab will not be linear. However, it is specified in BS8110 that the horizontal shear force due to ultimate loads should be taken as the total compression (or tension) force in the slab, where the interface is in the tension zone of the member. Where the interface is in the compression zone, the horizontal shear force should be taken as the compression force in that part of the zone which is above the interface. In either case, the average horizontal shear stress is determined by dividing the horizontal shear force by the area obtained by multiplying the contact width by the member length between the point of maximum ultimate sagging or hogging bending moment and the point of zero bending moment. For a simply supported beam carrying a uniformly distributed load, this length is clearly one-half of the span. The average horizontal shear stress is then distributed along the length of the interface in proportion to the vertical ultimate shear force diagram, to give the horizontal shear stress at any point along the interface. For a uniformly distributed load, this stress distribution would be the linear one shown in Fig. 10.10(b).

EXAMPLE 10.4   ■ ■

Determine the maximum horizontal shear stress for the simply supported composite section shown in Fig. 10.11 which spans 20 m. Assume that at the ultimate limit state the neutral axis of the composite section lies below the slab everywhere along the beam, and that f_{cu} for the *in situ* concrete is 30 N/mm². At midspan the total force in the slab

$$= 0.45 \times 30 \times 1500 \times 180 \times 10^{-3}$$
$$= 3645 \, \text{kN}.$$

Thus the average horizontal shear stress at the interface is given by

$$v_{h,avge} = \frac{3645 \times 10^3}{10.0 \times 10^3 \times 300}$$
$$= 1.22 \, \text{N/mm}^2.$$

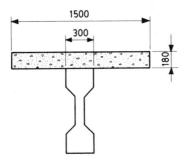

Fig. 10.11

The average horizontal shear stress must now be distributed along the length of the beam according to the shape of the ultimate shear force diagram. In this example, this is triangular and the maximum horizontal shear stress is given by

$$v_{h,max} = 2 \times 1.22$$
$$= 2.44 \, \text{N/mm}^2.$$

■ ■

The allowable ultimate horizontal shear stress is given in Table 10.2, from which it can be seen that the allowable stress is dependent on the surface texture of the beam at the interface, and also on the grade of the concrete. Providing nominal links greatly enhances the horizontal shear resistance, and this is reflected in Table 10.2.

Table 10.2 Allowable ultimate horizontal shear stresses (N/mm^2) at interface.

		Grade of in situ concrete		
Precast unit	*Surface type*	25	30	40 and over
Without links	As-cast or as-extruded	0.4	0.55	0.65
	Brushed, screeded or rough-tamped	0.6	0.65	0.75
	Washed to remove laitance or treated with retarder and cleaned	0.7	0.75	0.80
With nominal links projecting into *in situ* concrete	As-cast or as-extruded	1.2	1.8	2.0
	Brushed, screeded or rough-tamped	1.8	2.0	2.2
	Washed to remove laitance or treated with retarder and cleaned	2.1	2.2	2.5

In the above example, assuming that nominal links are provided in the beam and that the top surface is left as-cast, the allowable horizontal shear stress is 1.8 N/mm². All the horizontal shear force must now be carried by reinforcement anchored either side of the interface.

The amount of steel required, A_h, is given by

$$A_h = 1000 b_c v_h / (0.87 f_{yv})$$

where b_c is the breadth of the contact area. Thus, if f_{yv} for the links is 250 N/mm²,

$$A_h = 1000 \times 300 \times 2.44 / (0.87 \times 250)$$
$$= 3366 \, \text{mm}^2/\text{m}.$$

If R16 links are used at 115 mm centres, the total cross-sectional area is 3497 mm²/m.

This spacing is the maximum that is required near the supports and may be increased as the horizontal shear stress reduces away from the supports. Eventually, only nominal reinforcement is required and this may be determined as 0.15% of the contact area. The spacing of the links in T-sections should neither exceed four times the minimum thickness of the slab, not 600 mm. For the composite section in Example 10.4, the area of the nominal links is given by

$$A_h = (0.15/100) \times 1000 \times 300$$
$$= 450 \, \text{mm}^2/\text{m},$$

and the spacing required for R16 links is 890 mm. The maximum spacing, however, is the lesser of $4 \times 180 = 720$ mm, or 600 mm, and so the nominal links would be R16 at 600 mm centres. Alternative arrangements for the anchorage of the links in the slab are shown in Fig. 10.12.

For composite sections of the type shown in Fig. 10.5, the horizontal shear stress between the slab and the top of the beam may be determined using the

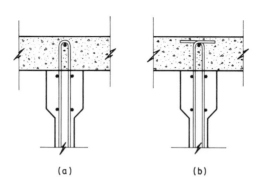

(a)   (b)

Fig. 10.12 Horizontal shear connection.

method described above. However, it is not necessary to check the shear stresses down the sides of the beam and along its lower flange, since these will generally be satisfactory if adequate provision has been made for horizontal shear resistance at the top of the beam. Further information on horizontal shear resistance may be found in Fédération Internationale de la Precontrainte (1982).

10.5 Vertical shear

As with the flexural strength of composite sections, the vertical shear resistance must be checked at two stages, one for the beam carrying the weight of the slab, and one for the composite section under service load. For the first stage, the shear resistance may be determined using the methods described in Chapter 7, and these may be adapted to check the shear resistance of the composite section.

EXAMPLE 10.5 ■ ■

Determine the shear resistance of the composite beam shown in Example 10.2, assuming the beam is uncracked in flexure at the ultimate limit state.

The ultimate shear force diagrams for the two stages to be considered are shown in Fig. 10.13.

For the first stage, where the beam is supporting the weight of the slab, the prestress f_{cp} at the centroid of the beam is $7.76 \, \text{N/mm}^2$. The maximum shear stress in the section is given by Equation 7.1, where, for the centroid, $A\bar{y} = 15.51 \times 10^6 \, \text{mm}^3$.

$$f_s = \frac{106.1 \times 10^3 \times 15.51 \times 10^6}{7.78 \times 10^9 \times 105}$$
$$= 2.01 \, \text{N/mm}^2.$$

The allowable principal tensile stress is $0.24 \times 50^{1/2} = 1.70 \, \text{N/mm}^2$. The un-

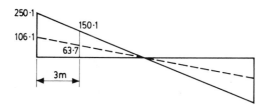

Fig. 10.13 Ultimate shear forces for beam in Example 10.2 (kN).

cracked shear resistance is given by Equation 7.6:

$$V_{co} = 0.67 \times 105 \times 695 \,(1.70^2 + 0.8 \times 1.70 \times 7.76)^{1/2} \times 10^{-3}$$
$$= 179.3 \,\text{kN}.$$

The uncracked shear resistance is thus satisfactory.

For the second stage, the total shear stress distribution is made up from that due to the first stage and that due to composite action under the imposed load. The shear stress distribution in the composite section, which is rectangular, is also given by Equation 7.1, with a maximum at the centroid of the composite section given by

$$f_{s,max} = \frac{(250.1 - 106.1) \times 10^3}{0.67 \times 500 \times 845}$$
$$= 0.51 \,\text{N/mm}^2.$$

The shear stress distributions within the beam at the two stages, together with that in the slab, are shown in Fig. 10.14. This shows that the maximum shear stresses in each case occur at different points in the section, and, furthermore, it can be shown that the point of maximum principal stress does not occur at the point of maximum shear stress. However, it is sufficiently accurate to base the shear resistance of the uncracked composite section on the principal tensile stress at the centroid of the composite section.

At this point the direct stress is found to be 7.97 N/mm². The principal tensile stress is then given by

$$f_{prt} = -7.97/2 + \tfrac{1}{2}[(7.97)^2 + 4(2.39)^2]^{1/2}$$
$$= 0.66 \,\text{N/mm}^2.$$

The uncracked composite section is thus adequate in shear. If the allowable

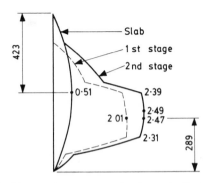

Fig. 10.14 Uncracked ultimate shear stresses for beam in Example 10.2 (N/mm²).

principal tensile stress had been exceeded, then the amount of shear reinforcement should be calculated using the methods described in Chapter 7.

An acceptable alternative method of finding the shear resistance of an uncracked beam is to find the principal tensile stress at the centroid of the composite section, with the shear stress at that point for each load stage given by

$$f_s = V/(0.67bh),$$

where V is the ultimate shear force, and b and h are the breadth and overall depth, respectively, for the appropriate section.

No consideration is usually given to whether the infill concrete is cracked. The adjacent precast beams provide restraint against cracking and generally it is satisfactory to check the principal tensile stress in the beam only.

It was shown in Chapter 7 that the shear resistance of prestressed concrete members depends on whether or not the section is cracked in flexure at the ultimate limit state, and that, generally, the lesser of the two shear resistances, V_{co} and V_{cr}, should be chosen. The method for determining V_{cr} for composite sections cracked in flexure follows that shown in Chapter 7. The difference is that the bending moment to give zero stress at the tensile face, M_0, is now given by

$$M_0 = M_d + (0.8f_{pt} - M_d/Z_{b,beam})Z_{b,comp} \tag{10.2}$$

where $Z_{b,beam}$ and $Z_{b,comp}$ are the lower fibre section moduli of the beam and composite section, respectively. f_{pt} is the prestress level at the lower fibre and M_d is the bending moment due to the weight of the beam and the slab. The shear resistance of the cracked section is then given by Equation 7.13.

EXAMPLE 10.6 ■ ■

Determine the shear resistance of the section in Example 10.2 assuming that it is cracked in flexure.

On the assumption that the tendons have a straight profile, at a section 3 m from a support:

$$f_{pt} = 7.76 + 11.82$$
$$= 19.58 \, \text{N/mm}^2;$$
$$M_d = (3.5 + 6.6) \times (3.0/2)(15.0 - 3.0)$$
$$= 181.8 \, \text{kN m}.$$

For the composite section,

$$Z_{b,comp} = 59.50 \times 10^6 \, \text{mm}^3.$$

Therefore, in Equation 10.2

$$M_0 = 181.8 + \left[(0.8 \times 19.58) - \frac{181.8 \times 10^6}{26.91 \times 10^6} \right] \times 59.50 \times 10^6 \times 10^{-6}$$
$$= 711.8 \, kN \, m.$$

The ultimate service load bending moment is

$$= \left[(3.5 + 6.6) \times 1.4 + 1.6 \times 12 \right] \times (3.0/2) \, (15.0 - 3.0)$$
$$= 600.1 \, kN \, m.$$

The ultimate shear force at 3 m from the support, from Fig. 10.13, is 150.1 kN. The allowable shear stress is found by taking b as the breadth of the precast beam rib, since the infill concrete is assumed to be cracked and to contribute little to the shear resistance of the composite section. The depth is taken as the depth of the beam plus the topping, since this region of *in situ* concrete does contribute to the shear resistance.
    Thus,

$$100 A_s/bd = (100 \times 1152)/(105 \times 755) = 1.45.$$

Therefore, from Table 7.2, $v_c = 0.71 \, N/mm^2$.
    The cracked shear resistance is given by Equation 7.13 as

$$V_{cr} = (1 - 0.55 \times 990/1770) \times 0.71 \times 105 \times 755 \times 10^{-3}$$
$$+ 711.8 \times 150.1/600.1$$
$$= 217.0 \, kN.$$

■ ■

The values of V_{cr} and V_{co} must be determined at several sections along the beam and the shear resistance diagram drawn, similar to Fig. 7.6. Any shear reinforcement required may be determined using the methods given in Chapter 7. For completeness, the shear resistance of the precast beam supporting its own weight and that of the slab should also be checked.

10.6 Deflections

The deflections of composite prestressed concrete members may be found using the same methods described in Chapter 6, depending on whether the beam is a Class 1, 2 or 3 member, but, as with the determination of stresses, account must be taken of the different types of section for each loading stage.

EXAMPLE 10.7   ■ ■

For the composite section in Example 10.2, determine the maximum deflections at the various load stages. Assume the eccentricity is constant at 200 mm for the middle third of the beam, reducing to zero at the supports.

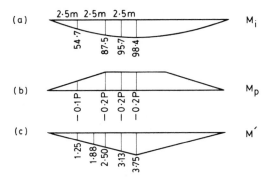

Fig. 10.15 Bending moment diagrams for beam in Example 10.7 (kN m).

(i) The bending moment distributions under the beam self weight, prestressing moment, and under a central unit point load are shown in Figs 10.15(a) 10.15(b) and 10.15(c) respectively. Using the virtual work method described in Chapter 6,

$$\delta_i = \int_0^L [M'(M_i + M_p)/EI]\,dx$$

$$= (2 \times 5/6EI)[4(54.7 - 0.1P)(1.25) + (87.5 - 0.2P)(2.50)]$$
$$+ (2 \times 2.5/6EI)[(87.5 - 0.2P)(2.50) + 4(95.7 - 0.2P)(3.13)$$
$$+ (98.4 - 0.2P)(3.75)].$$

With $\alpha P_i = 1250\,\text{kN}$,

$$\delta_i = -3685.1/EI.$$

For the beam, $I = 7.78 \times 10^{-3}\,\text{m}^4$ and $E = 28 \times 10^6\,\text{kN/m}^2$.

$$\therefore \delta_i = -\frac{3685.1}{7.78 \times 10^{-3} \times 28 \times 10^6}$$
$$= -0.0169\,\text{m}.$$

(ii) The bending moment distribution for the weight of the slab is shown in Fig. 10.16(a).

$$\delta_d - \delta_i = \int_0^L [M'(M_d - M_i)/EI]\,dx$$

$$= (2 \times 7.5/6EI)[4(1.88)(139.2) + (3.75)(185.6)]$$
$$= 4357.0/EI.$$

$$\therefore \delta_d - \delta_i = \frac{4357.0}{7.78 \times 10^{-3} \times 28 \times 10^6}$$
$$= 0.020\,\text{m}.$$

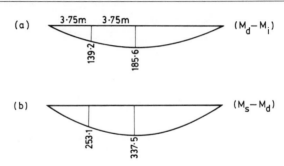

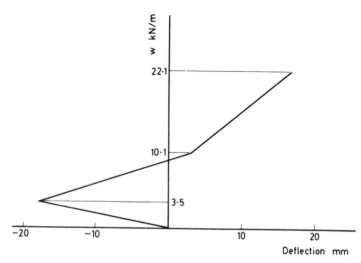

Fig. 10.16 Bending moment diagrams for beam in Example 10.7 (cont'd) (kN m).

Thus the total deflection under the weight of the slab is $-0.0169 + 0.020 = 0.0031$ m.

(iii) For deflections under the full service load, the composite second moment of area must be used. The bending moment distribution $(M_s - M_d)$ is shown in Fig. 10.16(b).

$$\delta_s - \delta_d = \int_0^L [M'(M_s - M_d)/EI]\,dx$$
$$= (2 \times 7.5/6EI)[4(1.88)(253.1) + (3.75)(337.5)]$$
$$= 7922.3/EI.$$

For the composite section, $I = 2.51 \times 10^{-2}\,\mathrm{m^4}$.

Fig. 10.17 Load–deflection curve for beam in Example 10.2.

Thus,

$$\delta_s - \delta_d = \frac{7922.3}{2.51 \times 10^{-2} \times 28 \times 10^6}$$
$$= 0.0113 \, \text{m}.$$

Under the full service load, the prestress force reduces to $1140 \, \text{kN}$, and so δ_i reduces to $-0.0145 \, \text{m}$.

The total deflection under the full service load is thus given by

$$\delta_s = -0.0145 + 0.020 + 0.0113$$
$$= 0.0168 \, \text{m}.$$

■ ■

A load–deflection curve for the composite section is shown in Fig. 10.17. This clearly shows the stiffening effect produced by the composite action of the slab and beam.

10.7 Differential movements

The fact that the slab of a composite member is usually cast at a much later stage than the beam means that most of the time-dependent effects of shrinkage of the slab take place with the section acting compositely. Most of the shrinkage of the beam will already have occurred by the time the slab is in place, and the movement due to the shrinkage of the slab will induce stresses throughout the whole of the composite section. The water content of the slab is often higher than that of the beam concrete, since a lower strength is required, and this aggravates the problem of differential shrinkage. These extra stresses, which occur even under zero applied load, are not insignificant and should be considered in design.

Both the slab and beam undergo creep deformations under load and, although some of the creep deformations in the beam may have taken place before casting of the slab, the level of compressive stress is higher in the beam, and so the creep deformations are larger. The differential creep deformations between the slab and beam set up stresses in the composite section which tend to reduce those set up by differential shrinkage.

A problem which is encountered, particularly in connection with bridge decks, is that of varying temperature across a composite section, although this may still be a problem in composite members used as roof structures. The hotter upper surface tends to expand more than the cooler lower surface and stresses are induced throughout the composite section.

A method for determining the stresses due to differential shrinkage will now be outlined, and this can be adapted to find the stresses due to differential creep and temperature movements.

Consider a composite member as shown in Fig. 10.18, where the slab is shown

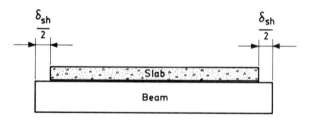

Fig. 10.18 Differential movements.

to have a free shrinkage movement of δ_{sh} relative to the beam. In reality this movement is restrained by the shear forces which are set up between the slab and beam, putting the slab into tension and the beam into compression. The magnitude of the tensile force in the slab is given by

$$T = \varepsilon_{sh} A_{c,slab} E_{c,slab}$$

where $A_{c,slab}$ and $E_{c,slab}$ are the cross-sectional area and modulus of elasticity of the slab, respectively and ε_{sh} is the free shrinkage strain of the slab concrete. The compressive force in the beam must be numerically equal to this tensile force.

In addition to the direct stresses described above, bending stresses are also introduced by restraint of the free differential shrinkage. In order to determine these stresses, the free bodies of the slab and beam are considered, as shown in Fig. 10.19. Initially, the slab can be regarded as having a force T applied through its centroid, so that its length is equal to that of the beam. There must be no net external force on the composite member due to differential shrinkage alone, so a

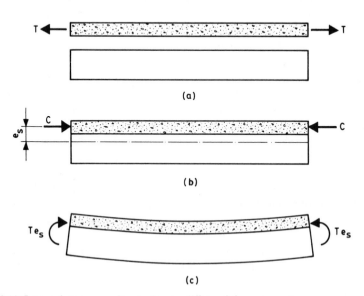

Fig. 10.19 Internal stress resultants due to differential movements.

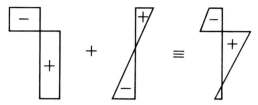

Fig. 10.20 Stresses due to differential movements.

pair of equal and opposite compressive forces must be applied to maintain equilibrium. However, these compressive forces act on the composite section and induce a bending moment at the ends of the member of magnitude Te_s, where e_s is the distance between the centroids of the slab and composite sections. The total stress distribution across the section is shown in Fig. 10.20.

EXAMPLE 10.8   ■■

For the composite section shown in Fig. 10.21, which has a span of 20 m, determine the direct stress distribution across the section if the slab undergoes a shrinkage strain of 100×10^{-6}. Take E_c for both regions of concrete as $28\,\text{kN/mm}^2$.

For the slab:

$$A_{c,\text{slab}} = 1500 \times 150 = 2.25 \times 10^5\,\text{mm}^2$$
$$\therefore\ T = 100 \times 10^{-6} \times 28 \times 10^3 \times 2.25 \times 10^5 \times 10^{-3}$$
$$= 630\,\text{kN}.$$

Thus the average stress in the slab $= -\dfrac{630 \times 10^3}{2.25 \times 10^5}$

$$= -2.80\,\text{N/mm}^2.$$

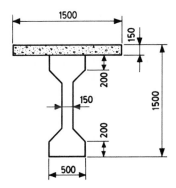

Fig. 10.21

For the beam,

$$A_{c,beam} = 2 \times 200 \times 500 + 150 \times 950 + 4 \times \frac{175^2}{2}$$

$$= 4.04 \times 10^5 \, mm^2.$$

Thus the average stress in the beam $= \dfrac{630 \times 10^3}{4.04 \times 10^5}$

$$= 1.56 \, N/mm^2.$$

The centroid of the composite section can be shown to lie 557 mm from the top of the slab. Thus the eccentricity of the slab centroid about the centroid of the composite section is $(557 - 75) = 482 \, mm$, and the moment about this centroid

$$= 630 \times 0.482$$
$$= 303.7 \, kN\,m.$$

For the composite section, $I_{comp} = 1.70 \times 10^{11} \, mm^4$.

Thus the bending stresses at the top of the slab, at the junction between slab and beam, and at the soffit of the beam are, respectively:

$$f_{t,slab} = \frac{303.7 \times 10^6 \times 557}{1.70 \times 10^{11}}$$

$$= 1.00 \, N/mm^2;$$

$$f_{b,slab} = f_{t,beam} = \frac{303.7 \times 10^6 \times 407}{1.70 \times 10^{11}}$$

$$= 0.73 \, N/mm^2;$$

$$f_{b,beam} = -\frac{303.7 \times 10^6 \times 943}{1.70 \times 10^{11}}$$

$$= -1.68 \, N/mm^2.$$

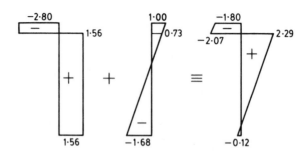

Fig. 10.22 Stress distribution for beam in Example 10.8 (N/mm²).

The resulting stress distribution is shown in Fig. 10.22. These stresses must be added to those due to the prestress force and applied load.

■ ■

10.8 Propping and continuity

All the composite sections considered so far have comprised beams which are simply supported. This is particularly useful where continued access is required beneath the structure throughout construction, such as where a bridge passes over a busy road or railway. However, if it is possible to provide a temporary intermediate support to the beam during construction, a considerable saving may be made, since the loading condition for the beam carrying the weight of the slab concrete has a large influence on its design.

A beam with a temporary central support is shown in Fig. 10.23. Initially the beam supports its own weight, with distribution of bending moments as shown in Fig. 10.24(a). Once the temporary support is in place and the slab concrete poured, the extra bending moments are as in Fig. 10.24(b). The stresses in the beam in these two cases are found using the beam section alone. When the concrete has hardened sufficiently, the temporary support is removed and the beam stresses are found for the slab load acting on the simply supported

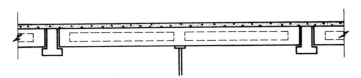

Fig. 10.23 Propping.

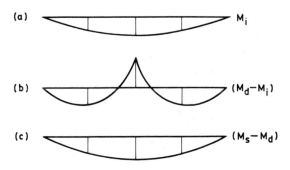

Fig. 10.24 Effects of propping.

composite section. Finally, the stresses induced by the imposed load bending moments, Fig. 10.24(c), acting on the composite section must be added. The final stresses in the beam will be less than if the beam had been unpropped, but the hogging bending moment induced in the beam when it is supporting the weight of the slab must be taken into consideration.

Clearly, the final stresses in the beam could be reduced further by introducing more temporary props at intermediate points in the span. The limiting case of this is to have the beam continuously propped. In this case, many of the advantages of using composite construction would be lost, but a good compromise may well be to have two or three intermediate supports, depending on the access required beneath the structure during construction.

Another extension of the basic form of composite construction is to join adjacent simply supported spans so that, under imposed load, they behave as a continuous structure. Two such simply supported sections are shown in Fig. 10.25, which support their own weight and that of the slab concrete. The slab extends across the top of the supports and is reinforced so that it can resist the tensile stresses which are set up there once the whole structure behaves continuously under the imposed load. The design of this region of the continuous member should be carried out as for a reinforced concrete member.

The continuous behaviour of the composite structure shown in Fig. 10.25 under imposed load will induce tensile stresses in the top of the prestressed concrete beams adjacent to the supports. These stresses should be limited to those in Table 5.3. In determining the ultimate strength of continuous members such as those shown in Fig. 10.25, the support sections should be considered as reinforced concrete sections. In the regions just beyond the bearing, the precompression in the concrete may be ignored over the transmission length of the tendons.

Another consideration in the use of continuous composite construction is the secondary bending moments set up at the supports due to creep and shrinkage in the adjacent spans. Long-term creep effects due to prestressing cause an upwards camber in the spans which induces a sagging bending moment at the support. Differential shrinkage and long-term creep effects due to the vertical

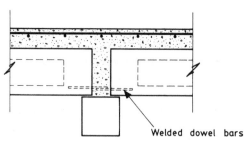

Fig. 10.25 Continuity.

load on the spans cause a downwards deflection in the adjacent spans, inducing a hogging bending moment at the supports. The overall effect is usually a net sagging bending moment, requiring reinforcement at the bottom of the support section, as shown in Fig. 10.25. Further information on the assessment of these secondary bending moments may be found in Clark (1978).

10.9 Design of composite members

The same considerations that were applied to the design of a prestressed concrete member in Chapter 9 may be applied when the member acts compositely with an *in situ* slab. However, there are now, in general, seven inequalities, since the stresses due to the dead load of the beam and slab, and the stress in the *in situ* concrete must also be considered. It is generally found that, of these seven stress conditions, the two most critical for determining the required prestress force and eccentricity are the upper fibre stress at transfer and the lower fibre stress at service load.

The minimum composite section size can be based on the stress conditions at the bottom fibre. When the beam is supporting its own weight, an inequality similar to 9.2(a) may be written:

$$\frac{\alpha P_i}{A_{c,beam}} + \frac{\alpha P_i e}{Z_{b,beam}} - \frac{M_i}{Z_{b,beam}} \leqslant f'_{max}, \tag{10.3}$$

where $A_{c,beam}$ and $Z_{b,beam}$ are the section properties of the beam. When the full service load is acting, the effect of the bending moment $(M_s - M_d)$ is found by using the composite section properties:

$$\frac{\beta P_i}{A_{c,beam}} + \frac{\beta P_i e}{Z_{b,beam}} - \frac{M_d}{Z_{b,beam}} - \frac{(M_s - M_d)}{Z_{b,comp}} \geqslant f_{min}, \tag{10.4}$$

where $Z_{b,comp}$ is the lower fibre section modulus for the composite section. Combining Inequalities 10.3 and 10.4 gives

$$Z_{b,comp} \geqslant \frac{\alpha(M_s - M_d)}{(\beta f'_{max} - \alpha f_{min}) + (1/Z_{b,beam})(\beta M_i - \alpha M_d)}. \tag{10.5}$$

The range for the prestress force required may be found for a given eccentricity from

$$P_i \geqslant \frac{(Z_{t,beam} f'_{min} - M_i)}{\alpha[(Z_{t,beam}/A_{c,beam}) - e]} \tag{10.6a}$$

$$P_i \geqslant Z_{b,beam} \frac{\{f_{min} + (M_s/Z_{b,comp}) + M_d[(1/Z_{beam}) - (1/Z_{b,comp})]\}}{\beta[(Z_{b,beam}/A_{c,beam}) + e]} \tag{10.6b}$$

Note that if the denominator in the first expression is negative, the inequality is reversed. Once the prestress force has been chosen, the limits to the eccentricity may be found from

$$e \leqslant \frac{Z_{t,beam}}{A_{c,beam}} + \frac{1}{\alpha P_i}(M_i - Z_{t,beam}f'_{min}), \tag{10.7a}$$

$$e \geqslant \frac{1}{\beta P_i}\left[Z_{b,beam}f_{min} + M_d\left(1 - \frac{Z_{b,beam}}{Z_{b,comp}} \right) + \frac{Z_{b,beam}}{Z_{b,comp}}M_s \right] - \frac{Z_{b,beam}}{A_{c,beam}}. \tag{10.7b}$$

Further information on composite construction design may be found in Bate and Bennett (1976).

References

Bate, S. S. C. and Bennett, E. W. (1976) *Design of Prestressed Concrete*, Wiley, New York.

Clark, L. A. (1978) *Concrete Bridge Design to BS5400*, Construction Press, London.

Fédération Internationale de la Précontrainte (1982) *Shear at the Interface of Precast and in situ Concrete*, Slough.

Chapter 11

INDETERMINATE STRUCTURES

11.1 Introduction

All of the prestressed concrete members so far considered have been statically determinate. This reflects the major use of prestressed concrete in building structures, since the most common type of prestressed concrete construction is in the form of simply supported beams. However, there are important applications of prestressed concrete in statically indeterminate structures. Many of the features of the analysis and design of these structures are similar to those used for statically determinate structures, which have been outlined in previous chapters. There are two important differences, however; namely, the introduction of secondary moments and behaviour at the ultimate limit state. These will be discussed in the following sections.

The most important application of prestressed concrete indeterminate structures is in the field of multi-span bridges. This is a specialized area of design and construction and is well beyond the scope of this book, but many good reference books on the subject may be found in the Bibliography.

In the field of building structures, continuous prestressed concrete beams are sometimes employed, but a more widespread use is in prestressed concrete flat slabs. The design of these will be discussed in detail in Chapter 12.

11.2 Secondary moments

It was shown in Chapter 1 that for a statically determinate prestressed concrete member the line of pressure in the concrete is coincident with the resultant force due to the prestressing tendons, provided that there is no applied axial load on the member. For statically indeterminate prestressed concrete structures, this is not necessarily the case. The prestress moment in a statically determinate member at any section is Pe, which is known as the *primary* prestress moment. In statically indeterminate structures, *secondary* or *parasitic* prestress moments may be introduced into the structure due to prestressing. Support reactions and shear forces will also be present in this case, even though there is no vertical load on the structure. The presence of these secondary moments involves extra work in the analysis and design of statically indeterminate prestressed concrete

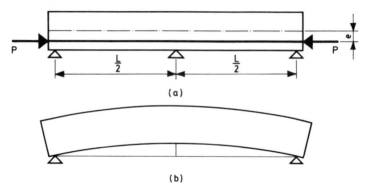

Fig. 11.1 Continuous prestressed concrete beam.

structures, although in nearly all other respects the design and analysis procedures outlined in the preceding chapters are applicable.

In order to understand how these secondary moments arise, consider a two-span continuous beam, as shown in Fig. 11.1(a), which has a constant prestressing force P acting at an eccentricity e.

If the central support were unable to restrain vertical upwards movement of the beam, the deflected shape of the beam due to the prestressing force would be as shown in Fig. 11.1(b). The beam is now effectively statically determinate and the prestress moment at any section would be the primary moment Pe (Fig. 11.2(a)). However, in practice the beam would be restrained at the central support, and in order to maintain compatibility of displacements at this position, a downwards reaction R must be applied at the support. The distribution of secondary moments induced in the beam by this reaction is shown in Fig. 11.2(b), while Fig. 11.2(c) shows the total distribution of the

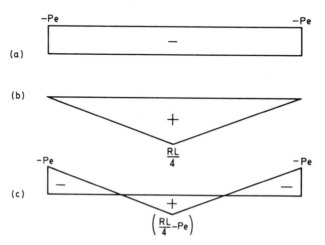

Fig. 11.2 Secondary moments.

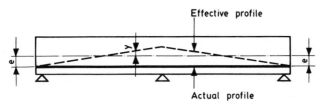

Effective profile

Actual profile

Fig. 11.3 Effective tendon profile.

moments along the beam. Note that the secondary moment diagram varies linearly between supports, since it is produced only by the reactions at the supports induced by prestressing.

The resulting prestress moment at any section shown in Fig. 11.2(c) may be written as Py, where y is some displacement. This may then be considered as the eccentricity of the resultant line of pressure in the concrete, and the locus of y along the beam may be considered as an effective tendon profile. The effective profile is obtained by raising or lowering the actual profile at interior supports, while keeping the basic shape of the profile constant, as shown in Fig. 11.3.

All the expressions given in Chapter 9 may be used for the design of statically indeterminate structures if, once the secondary moments have been determined, the actual eccentricity e is replaced by the effective eccentricity y.

EXAMPLE 11.1 ■■

A two-span continuous beam ABC has spans of 10 m and a prestress force of 1500 kN acting at a constant eccentricity of 300 mm. Determine the distribution of prestress moments along the beam.

On the assumption that there is no vertical restraint at the centre support, the beam is subjected to a pair of end-moments equal to Pe, that is, $1500 \times 0.3 = 450$ kN m. The midspan deflection of a beam subjected to a pair of end-moments M is given by

$$\delta_M = ML^2/8EI$$

where EI is the constant flexural stiffness of the beam and L is the span. Thus, for this example,

$$\delta_M = 450 \times 20^2/8EI = 22\,500/EI.$$

For a downward force R at the central support, the downward deflection at this point is given by

$$\delta_R = RL^3/48EI$$
$$= R \times 20^3/48EI = 166.7\,R/EI.$$

For compatibility of displacements at the central support, these two deflections must be numerically equal. Thus,

$$22\,500/EI = 166.7R/EI$$
$$\therefore R = 135\,\text{kN}.$$

The primary, secondary and total distributions of prestress moments along the beam are shown in Fig. 11.4(a)–(c), respectively, while the effective tendon profile is shown in Fig. 11.5.

An alternative method of analysis for the secondary moments is to consider the primary moment as a distributed applied moment on the structure. For the beam in this example, using the method of moment distribution, the fixed-end moments for each span may be found by considering the span as simply supported and with end-moments M_p and M_f applied as shown in Figs. 11.6(a) and (b). From symmetry, the fixed-end moments M_F at each end of the span must be equal. The rotations at each end of the span due to the combination of M_P and M_F are zero for a fixed-end condition. This rotation may be found conveniently using the virtual work method. The same simply supported span is shown in Fig. 11.6(c) with a unit moment applied to the left-hand end. The rotation at this end due to the moments M_P and M_F is then given by

$$1 \times \theta_A = \int_0^L [M'(M_p + M_F)/EI]\,\text{d}x.$$

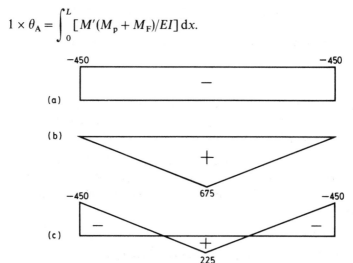

(a)

(b)

(c)

Fig. 11.4 Bending and prestress moments for beam in Example 11.1 (kN m).

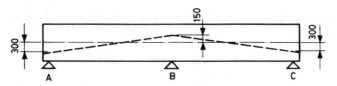

Fig. 11.5 Effective tendon profile for beam in Example 11.1.

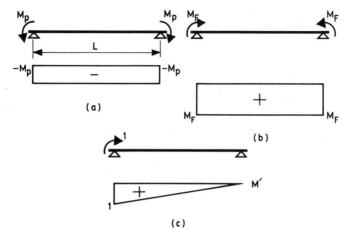

Fig. 11.6 Primary moment as distributed applied moment.

Table 11.1

	A	B		C
	AB	BA	BC	CB
D.F.		0.5	0.5	
F.E.M.	450	−450	450	−450
	−450 ⟶ −225		225 ⟵ 450	
Total	0	−675	675	0

From Simpson's rule,

$$\theta_A = (L/6EI)[(-M_P + M_F)(1) + 4(-M_P + M_F)(\tfrac{1}{2})]$$
$$= (-M_P + M_F)L/2EI.$$

Since this rotation must be zero, $M_p = M_F$. In this example, $M_p = 450 \, \text{kN m}$ and so the fixed-end moments are $450 \, \text{kN m}$. The moment distribution is shown in Table 11.1, showing that the secondary moment at support B is $675 \, \text{kN m}$, as found previously. The resulting total distribution of prestress moments throughout the structure is shown in Fig. 11.4(c).

■ ■

For less simple tendon profiles, the primary moment diagram shown in Fig. 11.6(a) is found by plotting the ordinates Pe along the span. The two fixed-end moments at either end of the span will, in general, be unequal, and the condition of zero end-slope must be applied to each end to enable solution for the unknowns.

The straight tendon profile shown in Fig. 11.1 was used only to illustrate how secondary moments arise. In practice the profile in continuous members is

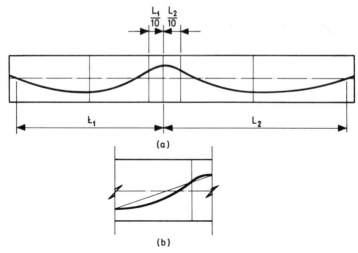

Fig. 11.7 Practical tendon profile.

determined according to the same underlying principle that is used for simply supported members, namely that the prestressing tendons are so positioned as to counteract any tension induced by the applied load. In continuous members, hogging support bending moments produce tension at the top surface, and so the eccentricity of the tendons is usually above the centroid at the supports. A typical tendon profile is shown in Fig. 11.7(a) and an enlarged detail of the tendon geometry near the support is shown in Fig. 11.7(b). The inflexion point for the profile is commonly taken as one-tenth of the span.

A useful method of determining the *total* prestress moments in an indeterminate structure is to analyse the structure under the equivalent loading applied to the concrete by the prestressing tendons. For a smoothly draped, or a sharply deflected, tendon a vertical force is exerted on the concrete member and the total distribution of prestress moments may be determined by any of the usual methods of structural analysis. The equivalent loading for the tendon profile shown in Fig. 11.7(a) is shown in Fig. 11.8.

For the straight tendon in Example 1.1, there is no vertical force exerted on the concrete, but there are end-moments as shown in Fig. 11.9. The moment distribution for the beam subjected to these end-moments is shown in Table 11.2 and the resulting distribution of total prestress moments is identical to that shown in Fig. 11.4(c).

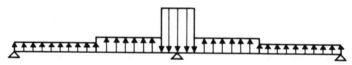

Fig. 11.8 Equivalent loading from tendons.

Fig. 11.9 Applied moments for straight tendon profile (kN m).

Table 11.2

	A	B		C
	AB	BA	BC	CB
D.F.		0.5	0.5	
F.E.M.	−450 →	→ −225	225 ←	← 450
Total	−450	−225	225	450

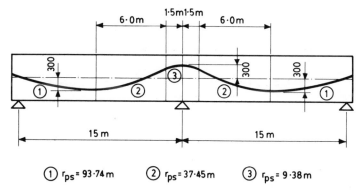

① $r_{ps} = 93 \cdot 74$ m   ② $r_{ps} = 37 \cdot 45$ m   ③ $r_{ps} = 9 \cdot 38$ m

Fig. 11.10

EXAMPLE 11.2   ■■

Determine the distribution of moments due to a prestress force of 1000 kN for the beam shown in Fig. 11.10.

The equivalent uniform vertical load exerted on the concrete is given by

$$w = P/r_{ps}.$$

Thus,

$w_1 = 1000/93.74 = 10.7 \, \text{kN/m}$

$w_2 = 1000/37.45 = 26.7 \, \text{kN/m}$

$w_3 = 1000/9.38 = 106.6 \, \text{kN/m}.$

The beam can now be analysed under the loading shown in Fig. 11.11.

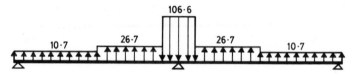

Fig. 11.11 Equivalent loading for beam in Example 11.2 (kN/m).

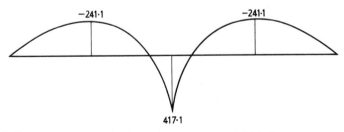

Fig. 11.12 Prestress moments for beam in Example 11.2 (kN m).

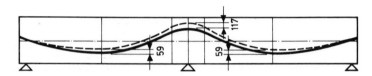

Fig. 11.13 Effective tendon profile for beam in Example 11.2.

The resulting distribution of total prestress moments is shown in Fig. 11.12, and the effective tendon profile in Fig. 11.13. Note that, once again, the effective tendon profile has been obtained by raising or lowering the actual profile at the supports, keeping the shape constant between the supports.

■ ■

The profile shown in Example 11.2 gives rise to an equivalent uniform load. If there is a sharp change in curvature, then the equivalent force on the concrete member is concentrated, as described in Chapter 1.

Once the secondary moments have been determined, the total stresses at the serviceability limit state may be found by adding the total prestress moment to the applied load bending moment. In analysing the structure for applied loads, the following combinations of load will be sufficient for most structures: (i) all spans loaded with full service load; (ii) alternate spans loaded with full service load and all other spans loaded with dead load only.

For framed structures, the axial shortening in the beams caused by prestressing induces horizontal reactions in the structure and gives rise to what are often called *tertiary* prestress moments. These must be considered in addition to the primary and secondary moments.

11.3 Linear transformation and concordancy

It was shown in the previous section that the line of pressure can be obtained from the actual tendon profile by raising or lowering the profile at interior supports while keeping the same basic shape between the ends of the member. This is an example of a *linear transformation* of a profile, since the amount by which the profile at any point is raised or lowered is directly proportional to the distance of that point from the interior supports which are adjusted.

Consider the beam shown in Fig. 11.14(a) and (b). The profile in Fig. 11.14(b) is a linear transformation of that in Fig. 11.14(a). The equivalent loads on the concrete in the two cases are shown in Fig. 11.15(a) and (b). The equivalent loads within the spans are the same, although the different inclinations of the tendons at the end supports give rise to different vertical forces there. Since the loads within the spans are the same, the total distribution of prestress moments must be equal, although the primary and secondary moments and the support reactions in each case would vary. The lines of pressure in each case must thus be equal.

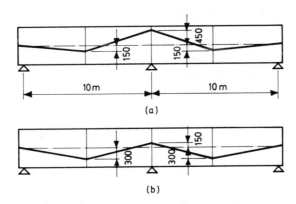

Fig. 11.14 Linear transformation.

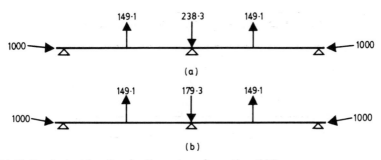

Fig. 11.15 Equivalent loading for linear transformation (kN).

In Fig. 11.14 the tendon profile has been shown with a sharp change of curvature at the central support, for simplicity. In practice the profile would be more like that shown in Fig. 11.7. For this profile, a linear transformation would slightly alter the inflexion point between the two curved regions and similarly affect the equivalent loads within the span. A linear transformation of such a profile, in theory, would thus cause a change in the total distribution of moments due to the prestress force, but in practice this change is very small and is usually ignored.

A general rule can now be stated, that if a tendon profile undergoes a linear transformation, the line of pressure in the concrete remains constant. This property is particularly useful in modifying tendon profiles when the limits to the cable zone, determined from Inequalities 9.6(a)–(d), do not permit a practical tendon profile. A profile can be chosen to lie within the theoretical cable zone, and a linear transformation performed to bring the actual profile into a more practical location.

If the eccentricity of the straight tendon profile shown in Fig. 11.1(a) is gradually reduced, so also will the free upward movement of the member at the central support position become smaller. The magnitude of the reaction required to maintain the beam in contact with the support will also be lessened, and this implies that the secondary moments will reduce. In the limiting case, the eccentricity becomes zero, and the beam is concentrically prestressed. The central support reaction and the secondary moments are then also zero. The total prestress moment in the beam at every section will be equal to the primary moment Pe, and the line of pressure will be everywhere coincident with the tendon profile.

Any tendon profile in a prestressed concrete member that has this property is known as a *concordant* profile. All profiles in statically determinate members are thus concordant, but in statically indeterminate members most profiles are non-concordant. For any given member, there can be many different basic profiles which are concordant.

In the design of a statically indeterminate prestressed concrete member, it is not necessary to ensure that the chosen profile is concordant, although this

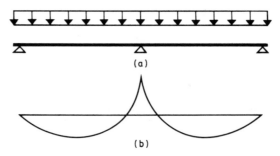

Fig. 11.16 Continuous beam with UDL.

simplifies the calculations. In practice it is found that concordant profiles are not the most economical, but in the design process it is quite useful to start with a concordant profile and then to modify it as necessary.

Consider now the continuous beam shown in Fig. 11.16(a) with a uniform load on each span. The distribution of bending moments is shown in Fig. 11.16(b). If tendons are fixed in the beam according to a profile determined from Fig. 11.16(b), then the equivalent load on the beam from the tendons must be of the form shown in Fig. 11.16(a), since any elastic bending moment distribution within a given structure can only correspond to one distribution of applied loads. The distribution of prestress moments within the member must therefore be similar to that shown in Fig. 11.16(b). Since this distribution of moments is consistent with zero vertical deflection at the central support, this tendon profile must be concordant.

A general rule thus emerges for determining concordant profiles, that the bending moment diagram for any given loading on a structure yields a concordant profile.

EXAMPLE 11.3   ■ ■

Determine a concordant profile for the beam shown in Fig. 11.17, using a prestress force of 500 kN.

A uniform load of 12 kN/m will be used to find a concordant profile. The bending moment diagram for this loading is shown in Fig. 11.18(a). The concordant tendon profile is obtained by dividing the ordinates of the bending moment diagram in Fig. 11.18(a) by the prestress force. The resultant profile is shown in Fig. 11.18(b).

■ ■

This profile is just one such concordant profile, since any linear transformation of it will not alter the position of the pressure line in the concrete and thus the profile will remain concordant. The design could then proceed exactly as for a statically determinate structure, since, provided the chosen tendon profile is a linear transformation of that shown in Fig. 11.18(b), the secondary moments will be zero.

The above method of finding a concordant profile is strictly valid only if the prestress force is constant along the member. In practice the prestress force varies, and the concordant profile should be obtained by dividing the ordinate

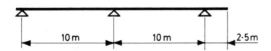

Fig. 11.17

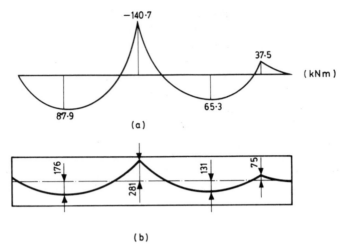

(a)

(b)

Fig. 11.18 Concordant profile for beam in Example 11.3.

of the bending moment diagram in Fig. 11.18(a) at any section by the effective prestress force at that section.

11.4 Ultimate strength behaviour

The analysis of prestressed concrete members at the ultimate limit state outlined in Chapter 5 is sufficient for statically determinate structures, since, for these structures, once the ultimate moment of resistance has been reached at any section, a mechanism is formed and the structure cannot support any more load.

The situation is different, however, for statically indeterminate structures. As the applied load is increased, the ultimate moment of resistance will be reached at some point in the structure, but in this case a mechanism will not form. Provided that the member at this point will allow rotation to take place at the *plastic hinge* which has formed, additional load can be carried by the structure, which effectively redistributes the load to less heavily-loaded regions until sufficient plastic hinges have formed to produce a mechanism. This description of the plastic analysis of prestressed concrete structures is equally applicable to steel, timber or reinforced concrete structures, and the general background and details of the theory may be found in Coates, Coutie and Kong (1980). The full methods of plastic analysis may be used for prestressed concrete structures, but there are important limitations imposed in BS8110 on the amount of rotation that takes place at a given section once a plastic hinge has formed there.

Consider the two-span continuous beam shown in Fig. 11.19(a), which is subjected to a total ultimate uniform load of 20 kN/m. An elastic analysis of the structure would give the bending moment distribution shown in Fig. 11.19(b). If the ultimate moment of resistance of the beam at the central support is 200 kN m, then the bending moment distribution of Fig. 11.19(b) can never be

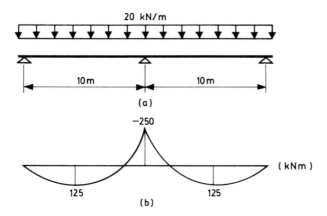

Fig. 11.19

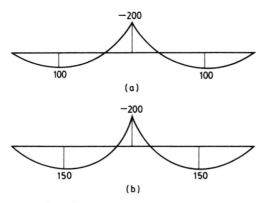

Fig. 11.20 Moment redistribution.

achieved. At a load of 16 kN/m, the bending moment diagram would be as shown in Fig. 11.20(a). As the uniform load is increased from 16 kN/m to 20 kN/m, the bending moment at the central support is assumed to remain constant at 200 kN m, since a plastic hinge has formed there. It order to maintain equilibrium, the bending moment diagram becomes that shown in Fig. 11.20(b). Comparison of this with Fig. 11.19(b) shows that the elastic bending moment of 250 kN m has been *redistributed* by 20% to give the value of 200 kN m in Fig. 11.20(b). However, the ultimate moment of resistance to be provided at the midspan sections has now increased to 150 kN m.

Moment redistribution may also be applied to the midspan sections. In this case it is the moment of resistance at the supports which must be increased to maintain equilibrium.

The 20% redistribution shown in the above example is the maximum permitted in BS8110. However, in practice the actual amount of redistribution

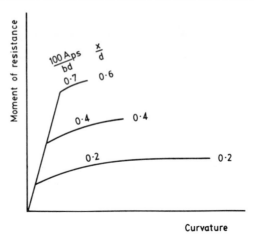

Fig. 11.21 Moment–curvature relationships.

permitted for a given section may be less than this figure. This is because of the fact that, for a statically indeterminate structure to resist an increase in load once the first plastic hinge has formed, rotation must take place at that hinge. The ability of a prestressed concrete member to undergo the required rotation once the ultimate moment of resistance has been reached is dependent on the position of the neutral axis within the section. Typical plots of moment–curvature for a given rectangular prestressed concrete section with varying amounts of prestressing steel are shown in Fig. 11.21. Each curve also corresponds to a different location of the neutral axis at the ultimate moment of resistance. For positions of the neutral axis high in the section, the amount of rotation that can take place after the plastic hinge has formed is much greater than for positions of the neutral axis lower in the section.

The amount of redistribution allowed is linked directly in BS8110 to the position of the neutral axis by the relationship

$$x \leqslant (\beta_b - 0.5)d, \tag{11.1}$$

where d is the depth to the centroid of the prestressing steel and β_b is the ratio of the ultimate bending moments at the section after and before redistribution. Thus, for the maximum value of redistribution of 20%, x must be less than $0.3d$. For most sections, this will mean that the member is under-reinforced, and such sections exhibit ductile behaviour, exactly the sort of behaviour necessary for moment redistribution to take place.

In assessing the elastic distribution of bending moments throughout the structure, the following load cases should be considered: (i) all spans loaded with $(1.4 \times$ dead load $+ 1.6 \times$ imposed load); (ii) alternate spans loaded with $(1.4 \times$ dead load $+ 1.6 \times$ imposed load) and all other spans loaded with $1.0 \times$ dead load.

Strictly according to plastic theory, the ultimate strength of a prestressed concrete structure is independent of any secondary moments induced by prestressing. However, in order to determine the limit to the neutral axis depth at a given section using Equation 11.1 the ultimate bending moments before redistribution should include any secondary moments present. The value of γ_f for these moments should be 1.0.

EXAMPLE 11.4 ■ ■

The two-span continuous beam shown in Fig. 11.22 has prestressing tendons with a total A_{ps} of 160 mm² and a constant effective prestress force of 180 kN. Determine the maximum ultimate uniform load, including self weight, that the beam can support. Assume that $f_{cu} = 40$ N/mm² and $f_{pu} = 1850$ N/mm².

At the central support,

$$100 A_{ps}/bd = (100 \times 160)/(200 \times 325) = 0.25;$$
$$100 f_{pe}/f_{pu} = (100 \times 180 \times 10^3)/(160 \times 1850) = 60.8.$$

From the design chart shown in Fig. 5.18, $M_u/bd^2 = 3.5$.

$$M_u = 3.5 \times 200 \times 325^2 \times 10^{-6}$$
$$= 73.9 \text{ kN m.}$$

Also from Fig. 5.18, x/d is approximately 0.26, by interpolation. Thus the maximum value of β_b allowed for the support section is given by

$$0.26 = \beta_b - 0.5;$$
$$\therefore \beta_b = 0.76.$$

For a uniform load w over the full beam, the maximum elastic support moment is $wL^2/8$, or $18.0\,w$. It can be shown that there is a sagging secondary moment of 9.0 kN m. Thus, with no redistribution, $w = 4.6$ kN/m. If 76% moment redistribution is allowed, then

$$0.76(-18.0w + 9.0) = -73.9;$$
$$\therefore w = 5.9 \text{ kN/m.}$$

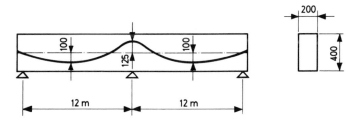

Fig. 11.22

For a support bending moment of 73.9 kN m, the corresponding midspan bending moment for this uniform load is 69.3 kN m. For the midspan sections,

$$100A_{ps}/bd = 100 \times 160/(200 \times 300) = 0.27:$$
$$100f_{pe}/f_{pu} = 60.8.$$

Thus, from Fig. 5.18,

$$M_u = 3.8 \times 200 \times 300^2 \times 10^{-6}$$
$$= 68.4 \text{ kN m}.$$

The ultimate moment of resistance at the midspans is thus inadequate, and a small amount of untensioned reinforcement is required in order for the beam to support the ultimate load of 5.9 kN/m.

■■

Reference

Coates, R. C., Coutie, M. G. and Kong, F. K. (1980) *Structural Analysis*, Nelson, Walton-on-Thames.

Chapter 12

PRESTRESSED FLAT SLABS

12.1 Introduction

The application of post-tensioned concrete to flat slab construction originated in the USA, and is now also widely used in Australia, Europe and the Far East. Although its use in the UK has been uncommon until recently, the economic advantages of this form of construction are now more widely appreciated. Most examples utilize uniform-depth slabs with draped tendons, but an interesting development is the use of variable-depth slabs with straight tendons. Extra shear strength is often provided around the columns in the form of a thickening, or drop.

No design guidance is given in BS8110 regarding prestressed concrete slabs. Design should, therefore, be based on the recommendations given in the Concrete Society (1979) Technical Report TR17. Many of the clauses in TR17 are similar to those in BS8110; only those which differ significantly will be outlined here.

Both bonded and unbonded tendons are employed in prestressed concrete flat slabs; the merits of each were discussed in Chapter 5. It is interesting to note that in the USA and the Far East, unbonded tendons are mainly used, while in Australia bonded construction is the norm. The choice would seem to depend on the local economics and design philosophy.

12.2 Two-way load balancing

The example given in Chapter 9 of a bridge deck slab was essentially a very wide simply supported beam. With a prestressed concrete slab simply supported on four edges, however, the situation is very different, since the structure is now highly indeterminate. A rectangular slab is shown in Fig. 12.1, supported on walls along each edge and prestressed with sets of uniformly spaced parabolic tendons in each direction.

For given tendon profiles and prestress forces in each direction, the slab could be analysed by theory of elasticity to give the distribution of stresses within the slab. This would be a complex problem, however, and an alternative method, based on the load balancing principle, is found to be very useful. The curved tendons exert an upwards force on the slab and, by suitable choice of profile and

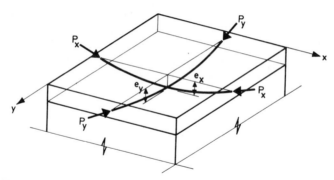

Fig. 12.1 Two-way prestressed concrete slab.

prestress force, any given load can be balanced to give zero deflections at every point in the slab.

If the prestress forces in the *x* and *y* directions are, respectively, P_x and P_y per unit length, and the midspan eccentricities are e_x and e_y, respectively, then since the vertical forces due to the tendons in each direction are additive at any point in the slab, the total upward uniform load on the slab is given by

$$w = 8P_x e_x/L_x^2 + 8P_y e_y/L_y^2. \tag{12.1}$$

Since the tendons must have a minimum spacing between them, the stress distribution within the slab will not be exactly uniform, but in practice it would be reasonably so.

Prestressed concrete slabs such as that shown in Fig. 12.1 are rarely found in practice, and the more common form is the flat slab, supported only on columns, with no intermediate beams. A prestressed concrete flat slab with irregularly spaced columns is shown in Fig. 12.2. Consider the area of the slab bounded by gridlines A, B, 1 and 3. *Primary* parabolic tendons are placed evenly between

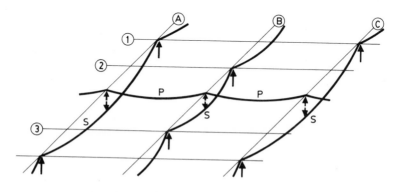

P = Primary tendon   S = Secondary tendon

Fig. 12.2 Transfer of load through tendons.

gridlines A and B and, by suitable adjustment of the profile and prestress force, the upward force from these tendons can be made to balance the applied load. Along gridlines A and B there will be a downward force exerted on the slab, due to the inclination of the tendons along these gridlines. For the slab shown in Fig. 12.1, these downward forces pass directly into the supporting walls, but in a flat slab there are no supporting beams or walls and some other means of resisting the forces is necessary.

The downward forces are resisted by *secondary* parabolic tendons along gridlines A and B. The upwards forces from these balance the downwards forces from the primary tendons and the downwards forces from the secondary tendons pass directly into the columns. The uniform applied load on the slab has thus passed into the columns through the system of prestressing tendons, leaving the slab level and in a uniform state of compression. In practice, the forces in all the tendons are not constant and do not balance each other exactly, but the actual state of stress within the slab will be reasonably uniform.

The slab shown in Fig. 12.2 will resist any additional imposed load in much the same way as would a reinforced concrete flat slab, and the same analysis methods are applicable. Since this analysis need only be carried out for a small percentage, usually, of the total service load, any inherent inaccuracies in the method are not significant.

In practice, the sharp changes of curvature in the tendons shown in Fig. 12.2

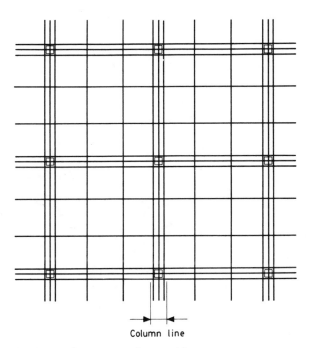

Column line

Fig. 12.3 Tendon layout for square column grid.

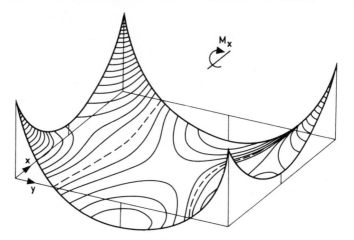

Fig. 12.4 Bending moment distribution in flat slab panel due to applied loads.

are avoided by adopting smooth reverses in curvature, as shown in Chapter 11. A suitable layout of tendons within a flat slab with regularly spaced columns in each direction is shown in Fig. 12.3. The secondary tendons are shown extending throughout *column lines*, which are equal in width to the width of the column plus a half-slab depth either side of the column faces.

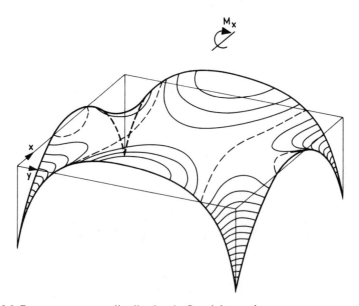

Fig. 12.5 Prestress moment distribution in flat slab panel.

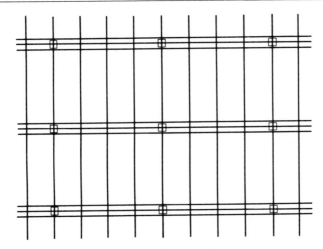

Fig. 12.6 Tendon layout for rectangular column grid.

The distribution of bending moments in an interior panel of a flat slab, subjected to a uniform load, is shown in Fig. 12.4, and the distribution of prestress moments obtained from a layout of tendons similar to that shown in Fig. 12.3 is shown in Fig. 12.5 (Figs 12.4 and 12.5 are adapted from Birkenmaier, Welbergen and Winkler (1986)). The net bending moment distribution is thus very even, with small peaks along the column lines.

An alternative pattern of tendons which is often used is shown in Fig. 12.6, based on the distribution of primary and secondary tendons shown in Fig. 12.2.

12.3 Equivalent frame analysis

The load balancing technique is very useful for estimating the prestress force required in each direction, but an analysis of the slab for the unbalanced loads must still be made. The analysis and design of reinforced concrete flat slabs with regularly-spaced columns has for many years been based on a method which divides the slab and columns into a series of equivalent frames in each direction. The distributions of bending moment and shear force may be determined by any of the available methods of structural analysis. The equivalent frame method has also been found to give acceptable results for prestressed concrete flat slabs and is one of those recommended in TR17. It can be used to find the distribution of moments due to both the prestress force and the full service load, and not just the unbalanced loads.

Each equivalent frame comprises columns and strips of slab at each floor level. The width of slab to be used to determine the beam stiffness is equal to the full width of the panels for vertical loading, while for horizontal loading, where

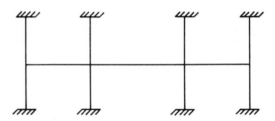

Fig. 12.7 Equivalent frame.

lateral stability is provided by the frame, one-half of the panel width should be used.

Each frame may be analysed as a whole, but, for vertical loads only, each strip of slab at a given floor level may be analysed as a separate frame, with columns above and below the slab assumed to be fixed in position and direction (Fig. 12.7). Where drops are provided around the columns, an equivalent slab thickness may be found for the support section of the slab, with the same second moment of area as the true section, and the overall beam stiffness modified accordingly. Most slab layouts have a regular column spacing in both directions, but for layouts such as that shown in Fig. 12.2 the equivalent frame method is not suitable, and in order to analyse such slabs numerical methods such as finite elements must be used.

If the equivalent frame method is used to determine the stresses at service load, the analysis should be carried out for the following load cases:

(i) All spans loaded with full service load.

(ii) Alternate spans loaded with full service load, all other spans loaded with dead load only.

If the frame provides lateral stability for the structure, the load cases considered must also include the wind load, with the respective partial factors of safety for dead, imposed and wind load described in Chapter 3.

As in the case of beams, if the load balancing technique is employed to estimate the prestress force required, it must be decided what proportion of the service load is to be balanced, and, as described earlier, a common criterion is to balance the permanent load. The stresses at service load should then be checked under the loading conditions stated above. Stresses at transfer should be checked assuming that only the dead load of the slab is acting, plus, perhaps, a small allowance for construction loads.

For a set of tendons which have been designed to balance a given applied load, Equation 12.1 indicates that the load is distributed to the supports by the tendons in each direction. However, in a flat slab, the 'supports' are themselves tendons, and in order to maintain equilibrium the total upwards force in *each* direction provided by the tendons along the column lines, and in the interval between them, must balance the *total* load on the panel. Thus, in using the load balancing technique to estimate the prestress force in each direction, prior to carrying out an equivalent frame analysis, the tendons are determined using the full load to be balanced.

In determining the maximum allowable drape of the tendons in each direction, it is important to remember that the tendons have a finite diameter and, where they cross, one layer must be detailed to lie on top of the other, thus reducing the effective depth of the former.

12.4 Design and detailing

Much of the design of prestressed concrete flat slabs is similar to that of beams outlined in the preceding chapters. Some of the differences from those methods are described here.

For preliminary estimation of the required slab depth, span/depth ratios in the range of 42 to 48, for floors and roofs, respectively, may be assumed. These values may be raised to 48 and 52, respectively, provided that the imposed load does not exceed $3\,kN/m^2$ and that serviceability criteria are satisfied. Waffle slabs are sometimes used, and the span/depth ratio for these should not exceed 35. Careful attention must be paid to the layout of tendons in these slabs, since they can only be placed in the ribs. Solid slabs with spans in excess of 10 m are likely to require drop panels in order to provide adequate shear resistance at the columns, and waffle slabs with smaller spans than this will require at least a solid section near the column.

The allowable concrete stresses at the various loading stages and locations

Table 12.1 Allowable concrete stresses in slabs.

		In tension	
Loading condition	In compression	With bonded reinforcement	Without bonded reinforcement
Maximum stress at transfer			
At positive moment locations	$0.33f_{ci}$	$0.45f_{ci}^{1/2}$	$0.15f_{ci}^{1/2}$
At negative moment locations	$0.24f_{ci}$	$0.45f_{ci}^{1/2}$	0
Maximum stress at service loads			
At positive moment locations	$0.33f_{cu}$	$0.45f_{cu}^{1/2}$	$0.15f_{cu}^{1/2}$
At negative moment locations	$0.24f_{cu}$	$0.45f_{cu}^{1/2}$	0

are shown in Table 12.1. These stresses are conservative, in order to allow for the fact that approximate analysis methods such as the equivalent frame method underestimate the maximum negative bending moments at the supports. The allowable tensile stresses in Table 12.1 relate to unbonded construction, and no account is taken of the better resistance to cracking of bonded construction. The stresses in Table 12.1 may be relaxed if a rigorous elastic analysis is performed. The allowable stresses given in Chapter 3 may then be used, provided that deflections and ultimate strength capacity are satisfied.

It is recommended in TR17 that a suitable layout of tendons will result if 50% or more of the total number of tendons in a panel, determined from an equivalent frame analysis, are placed in the column lines in each direction, and the rest of the tendons uniformly distributed, as shown in Fig. 12.3. In practice, however, the tendon layout shown in Fig. 12.6 is commonly employed, although the layout in Fig. 12.3 provides better punching shear resistance at the columns (see Section 12.6). Maximum spacing of tendons or groups of tendons should be six times the slab depth and the minimum spacing should be 75 mm for single tendons, and for grouped, unbonded tendons should be twice the width of the group. Cover requirements should be determined from Tables 3.4 and 3.6, and the depth of any granolithic screed on the slab may be taken into account when considering these requirements.

Untensioned reinforcement should be placed in the top of all slabs over the columns, equal in area to a minimum of 0.15% of the gross cross-section of the slab, based on a width equal to that of the column plus twice the slab depth on each side of the column. Bars should be at least 12 mm diameter, extend at least one-sixth of the span on either side of the column and have a maximum spacing of 300 mm. The combination of tendons and reinforcing bars in the region of the columns could lead to congestion, and careful attention should be paid to the detailing in this area.

For unbonded tendon construction, if the allowable concrete tensile stresses of $0.45f_{ci}^{1/2}$ or $0.45f_{cu}^{1/2}$ are to be used from Table 12.1, untensioned reinforcement must be provided. The stress distribution within the section over a support at the serviceability limit state may be assumed to be that shown in Fig. 12.8. The

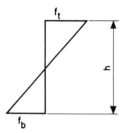

Fig. 12.8 Stress distribution at serviceability limit state.

untensioned reinforcement should be capable of resisting the total tensile force in the section, so that the area of steel required per unit length is given by

$$A_s = F_t/0.58 f_y,$$

where

$$F_t = f_t^2 . h/2(f_b + f_t). \tag{12.2}$$

Note that f_t in Equation 12.2 should be taken as positive.

Particular attention should be paid to the detailing of reinforcement around the anchorages, since the bursting forces may be high in the thin slab edge. Further information on the design and detailing of prestressed concrete flat slabs may be found in reports of the Concrete Society (1979, 1984).

EXAMPLE 12.1   ■■

A prestressed concrete flat slab warehouse floor has the layout shown in Fig. 12.9. The imposed load is $10 \, \text{kN/m}^2$. Determine a suitable slab depth and layout of prestressing tendons.

Since the floor is heavily loaded, a span/depth ratio of 38 will be chosen. The slab depth required is then $7.5/38 = 0.197 \, \text{m}$. An initial slab depth of 200 mm will be assumed.

Loading:

Slab self weight	$4.8 \, \text{kN/m}^2$,
Finishes + services	$1.5 \, \text{kN/m}^2$.

Material properties:

$$f_{cu} = 40 \, \text{N/mm}^2$$
$$f_{pu} = 1770 \, \text{N/mm}^2 \text{ for } 15.7 \, \text{mm super strand.}$$

Since the imposed load is large compared with the dead load, the load to be

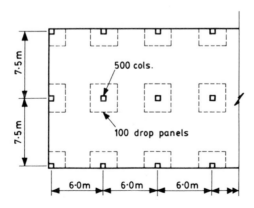

Fig. 12.9

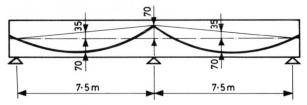

Fig. 12.10 Idealized tendon profile.

balanced will be taken as the dead load plus one-third of the imposed load, that is, $6.3 + (10.0/3) = 9.6 \, \text{kN/m}^2$.

Cover to the tendons is 20 mm, so that with the outside diameter of the sheathed tendons as 19 mm, the maximum eccentricity of the tendons = $100 - (20 + 19/2) = 70 \, \text{mm}$.

The preliminary design for the tendons in the transverse direction will only be presented here. The design for the longitudinal tendons is similar, however.

The idealized profile of the tendons is shown in Fig. 12.10. The drape of the tendons in each span is 105 mm, so that the required prestress force to balance $9.6 \, \text{kN/m}^2$ is given by

$$P_x = wL^2/8d_r$$
$$= (9.6 \times 7.5^2)/(8 \times 0.105) = 642.9 \, \text{kN/m}.$$

Assuming $\beta = 0.85$, the initial prestress force required is 756.3 kN/m.

For a 15.7 mm super strand, with $A_{ps} = 150 \, \text{mm}^2$, the initial prestress force is 185.9 kN, and thus the total number of tendons required per bay is 25.

The equivalent frame to be analysed is shown in Fig. 12.11. The effect of the drop panel is taken into account by considering an equivalent uniform slab section, Fig. 12.12(a). For the actual slab section, $I = 8.86 \times 10^9 \, \text{mm}^4$. Thus an effective slab depth, h_{eff} (Fig. 12.12(b)) is given by

$$h_{eff} = \left(\frac{8.86 \times 10^9 \times 12}{6 \times 10^3} \right)^{1/3}$$

$$= 261 \, \text{mm}.$$

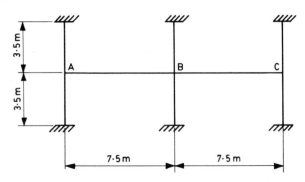

Fig. 12.11 Equivalent frame.

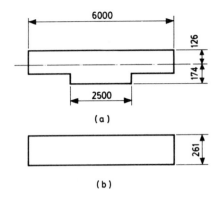

Fig. 12.12 Equivalent slab section at drop panel.

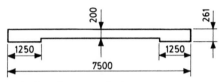

Fig. 12.13 Beam for equivalent frame analysis.

The beams in the equivalent frame of Fig. 12.11 thus have varying depth, as shown in Fig. 12.13, and using tables of coefficients for such members (Concrete Society, 1984), the following information necessary for the moment distribution analysis is obtained:

  beam stiffness $= 6.2EI/L$
  carry-over factor $= 0.6$;
  fixed-end moment $= 0.091\ wL^2$.

Thus,

$$k_{col} = (4 \times 500^4)/(12 \times 3500) = 5.95 \times 10^6;$$
$$k_{beam} = (6.2 \times 200^3 \times 6000)/(12 \times 7500) = 3.31 \times 10^6.$$
$$\therefore r_{AB} = 0.22$$

and

$$r_{BA} = 0.18.$$

Fixed-end moments:

  dead load $= 6.3 \times 6.0 \times 7.5^2 \times 0.091\ = 193\ \mathrm{kN\,m}.$

  imposed load $= 10.0 \times 6.0 \times 7.5^2 \times 0.091 = 307\ \mathrm{kN\,m}.$

The bending moment envelope for the load combinations specified previously is shown in Fig. 12.14.

The equivalent load from the tendons in each span

$$= 8 \times 0.85 \times 25 \times 185.9 \times 0.105/7.5^2$$
$$= 59.0\ \mathrm{kN/m}.$$

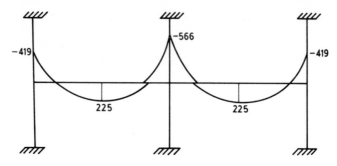

Fig. 12.14 Service load bending moment envelope (kN m).

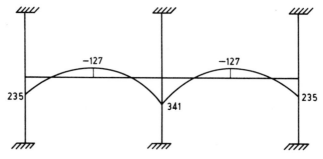

Fig. 12.15 Prestress moments (kN m).

The prestress moment diagram is shown in Fig. 12.15. Note that no account has been taken at this stage of the reversed curvature of the tendons that must take place over the column lines. The error thus introduced is small.

For support sections,

$Z_t = 8.86 \times 10^9/126 = 70.32 \times 10^6 \, \text{mm}^3;$

$Z_b = 8.86 \times 10^9/174 = 50.92 \times 10^6 \, \text{mm}^3;$

$A_c = 6000 \times 200 + 2500 \times 100 = 1.45 \times 10^6 \, \text{mm}^2.$

For midspan sections,

$Z_b = Z_t = 6000 \times 200^2/6 = 40.00 \times 10^6 \, \text{mm}^3;$

$A_c = 6000 \times 200 = 1.20 \times 10^6 \, \text{mm}^2.$

The resulting stresses at the critical sections are shown in Table 12.2, where M is the sum of the moments shown in Figs 12.14 and 12.15.

From Table 12.1, for $f_{cu} = 40 \, \text{N/mm}^2$, the values of f_{max} are 13.20 N/mm² and 9.60 N/mm² for midspan and support sections, respectively, while the corresponding values of f_{min} are $-2.85 \, \text{N/mm}^2$ for both locations, assuming there is sufficient untensioned reinforcement.

Table 12.2 Service load stresses for slab in Example 12.1.

		P/A_c (N/mm^2)	M/Z (N/mm^2)	Total (N/mm^2)
Outer	Top	2.72	−2.62	0.10
Columns	Bottom	2.72	3.61	6.33
Midspan	Top	3.29	2.45	5.74
	Bottom	3.29	−2.45	0.84
Inner	Top	2.72	−3.20	−0.48
Columns	Bottom	2.72	4.42	7.14

The stresses in Table 12.2 for the column sections of the slab have been determined for the slab depth at the drop panel. If the bending moments in the slab from Figs 12.14 and 12.15 are assumed to be uniformly distributed across the panel width, the stresses in the slab at the column sections away from the drop would be higher than those in Table 12.2. However, in practice, the bending moments in flat slabs are not uniform but have peaks over the columns, as shown in Fig. 12.4, and this is one of the reasons that 50% of the tendons in a panel are grouped in the column lines. The analysis of stresses described above is therefore only an approximation to the true state of stress in the slab, but has been found to give satisfactory results.

The amount of untensioned reinforcement required at the interior supports is given by

$$A_s = \frac{0.48^2 \times 300}{2(7.14 + 0.48)} \times \frac{1000}{0.58 \times 460}$$
$$= 17 \, mm^2/m.$$

A layer of A98 mesh would be adequate.

In order to check the transfer stresses, assume $\alpha = 0.95$. The bending moment distribution due to the prestress force and the dead load of the slab only is shown in Fig. 12.16, and the corresponding stresses are shown in Table 12.3.

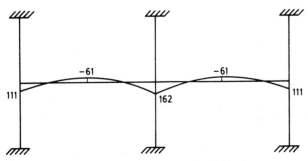

Fig. 12.16 Bending and prestress moments at transfer (kN m).

Table 12.3 Transfer stresses for slab in Example 12.1.

		P/A_c (N/mm^2)	M/Z (N/mm^2)	Total (N/mm^2)
Outer	Top	3.04	1.58	4.62
Columns	Bottom	3.04	-2.18	0.86
Midspan	Top	3.68	-1.53	2.15
	Bottom	3.68	1.53	5.21
Inner	Top	3.04	2.30	5.34
Columns	Bottom	3.04	-3.18	-0.14

For $f_{ci} = 30\,N/mm^2$, the values of f'_{max} are $9.90\,N/mm^2$ and $7.20\,N/mm^2$ for midspan and support sections, respectively, while f'_{min} is $-2.46\,N/mm^2$ for both sections. Thus, if untensioned reinforcement is provided top and bottom at the supports the stresses in Table 12.3 are satisfactory.

■ ■

The design of the slab so far presented is only preliminary. The tendon profiles must be revised to give a smooth change of curvature over the column lines and the distribution of prestress force along the tendons must be determined. A revised analysis of the moments in the slab due to the prestress force may then be carried out and the revised stresses checked against the allowable ones. With experience, the preliminary design should need very little amendment.

12.5 Ultimate strength

The equivalent frame method described in Section 12.3 may be used to analyse a prestressed concrete flat slab at the ultimate limit state, using the load combinations described in Chapter 11. However, in most cases it is sufficient to consider the single case of maximum design ultimate load on all spans, provided that the ratio of imposed load to dead load does not exceed 1.25 and that the imposed load does not exceed $5\,kN/m^2$. The support bending moments thus obtained, except those adjacent to cantilevers, should be redistributed by 20%.

The determination of the ultimate strength of the slab at the critical sections may be carried out according to the methods given in Chapter 5 for both bonded and unbonded tendons. If the ultimate strength capacity based on the tendons alone is insufficient, then additional untensioned reinforcement must be added. Most prestressed concrete slabs have small percentages of prestressing steel and are thus under-reinforced. Advantage can therefore usually be taken of moment redistribution, although this is limited in TR17 to 15% for negative moments. The limitations on neutral axis depth given in Chapter 11 must also be observed.

EXAMPLE 12.2

■ ■

Determine the ultimate strength capacity of the slab in Example 12.1.

$$\text{Ultimate uniform load} = 1.4 \times 6.0 \times 6.3 + 1.6 \times 6.0 \times 10.0$$
$$= 149 \, \text{kN/m}.$$

The envelope of bending moments for the load conditions specified in Chapter 11 is shown in Fig. 12.17. To this should be added the distribution of secondary moments.

The depth of the slab at the supports, and hence its ultimate strength, varies across the panel width due to the increase in slab depth at the drop. However, it is sufficient to check the ultimate strength of the support section based on the depth of the slab at the drop and the average bending moment and average area of tendons across the panel. The high concentration of actual ultimate bending moments over the columns will be compensated by the grouping of the tendons in the column lines.

For the support sections,

$$\text{Effective depth} = 300 - (20 + 19/2) = 270 \, \text{mm}.$$

$$\text{Area of prestressing tendons } A_{ps} = 625 \, \text{mm}^2 \text{ per metre width of slab.}$$

$$f_{pe} = 0.85 \times 0.7 \times 1770 = 1053 \, \text{N/mm}^2.$$

From Equation 5.4,

$$f_{pb} = 1053 + \frac{7000}{7500/270}\left(1 - 1.7 \times \frac{1770 \times 625}{40 \times 1000 \times 270}\right)$$

$$= 1261 \, \text{N/mm}^2 (< 0.7 f_{pu}).$$

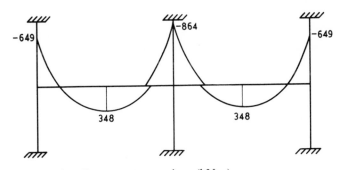

Fig. 12.17 Ultimate bending moment envelope (kN m).

Thus, in Equation 5.5, $x = 2.47 \times \dfrac{625 \times 1261}{40 \times 1000}$

$$= 49\,\text{mm}.$$

$$\therefore M_{\text{u}} = 1261 \times 625\,(270 - 0.45 \times 49) \times 10^{-6}$$

$$= 195.4\,\text{kN m/m}.$$

It can be shown that there is a sagging secondary moment of 64 kN m at the interior support. The maximum total bending moment at the interior support is thus 800 kN m, which represents an average bending moment across the section of 133.3 kN m/m.

For the midspan section,

effective depth $= 200 - (20 + 19/2) = 170\,\text{mm}$;

$$f_{\text{pb}} = 1053 + \frac{7000}{7500/170}\left(1 - \frac{1.7 \times 1770 \times 625}{40 \times 1000 \times 170}\right)$$

$$= 1168\,\text{N/mm}^2;$$

$$x = 2.47 \times \frac{625 \times 1168}{40 \times 1000}$$

$$= 45\,\text{mm}.$$

$$\therefore M_{\text{u}} = 1168 \times 625(170 - 0.45 \times 45) \times 10^{-6}$$

$$= 109.3\,\text{kN m/m}.$$

The average ultimate bending moment across the width of the panel at midspan, including the secondary moment, is 83.0 kN m/m, and thus the slab has ultimate strength capacity.

■ ■

12.6 Shear resistance

Wherever a concentrated force is applied to a slab, the shear forces local to the force are high. In flat slabs, where the loads are predominantly uniform, the areas where the shear resistance of the slab must be checked are around the columns, although if a heavy concentrated load is applied, the shear resistance around this must also be checked.

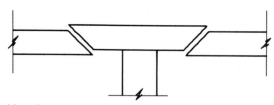

Fig. 12.18 Punching shear.

The determination of the shear resistance of prestressed concrete flat slabs is carried out in much the same way as for reinforced concrete flat slabs. The clauses in TR17 are based on those in CP110 for reinforced concrete slabs, but since these have been revised in BS8110, the design procedure presented here is based on the new code.

A section through a flat slab around a column is shown in Fig. 12.18 and illustrates the typical 'punching' mode of failure. The actual failure is modelled by assuming vertical failure zones around the column, shown in plan in Fig. 12.19. Each is defined by a critical perimeter, u, and the average shear stress, v, along this perimeter is defined by

$$v = V_t/(ud), \tag{12.3}$$

where V_t is the shear force along the perimeter and d is the effective depth of the prestressing tendons. The value of v determined from Equation 12.3 should be compared with the allowable stress, v_c, taken from Table 7.2. The maximum value of v should be $0.8f_{cu}^{1/2}$ or $5\,\mathrm{N/mm^2}$, whichever is less, based on the perimeter of the actual column.

In order to find the effective shear force at a column, the bending moments and shear forces should first be determined from an equivalent frame analysis based on the pattern loading conditions described in the previous section.

It is recognized that, where an appreciable bending moment is induced in a column, the distribution of shear forces within the slab is not uniform around the column. An effective shear force, V_{eff}, is used to allow for this effect, given by

$$V_{eff} = V_t[1 + (1.5M_t/V_tx_p)], \tag{12.4}$$

where V_t and M_t are, respectively, the support reaction and bending moment transmitted to the column by the slab, and x_p is the side of the critical perimeter parallel to the axis of bending. In lift slab construction, M_t should be taken as zero.

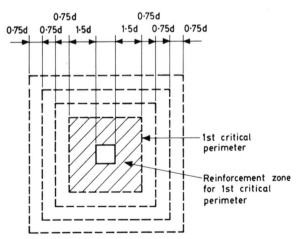

Fig. 12.19 Failure zones.

For approximately equal spans, and where the structure is braced against wind loads, V_{eff} may be taken as 1.15 V_t for internal columns where V_t is based on the case of maximum ultimate load applied to all adjacent panels. If the spans in each direction are substantially different, Equation 12.4 should be applied independently for each direction and the design checked for the worse case. For corner columns and for bending of edge columns about an exis parallel to the edge, V_{eff} may be taken as 1.25 V_t. For bending of edge columns about an axis perpendicular to the edge, V_{eff} may be taken as 1.4 V_t for approximately equal spans.

The value of v_c from Table 7.2 is dependent on the total tensile force crossing the critical perimeter and account should be taken of both prestressing tendons and reinforcing bars. The value of A_s in Table 7.2 should be taken as A_s^*, given by

$$A_s^* = A_s + (f_{pu}/460)A_{ps},$$ (12.5)

where A_{ps} and A_s are the areas of the prestressing tendons and reinforcing bars, respectively. If the tendon and bar arrangements in each direction are different, the value of v_c along each side of the critical perimeter should be considered separately and the contributions to shear resistance added. The value of A_s^* should not be taken as greater than 3.0% of bd.

The shear resistance at the first critical perimeter should be checked and, if v is found to exceed v_c, then either the slab thickness can be increased locally by providing a column head or a drop panel, or shear reinforcement placed within the slab. The shear resistance along successive critical perimeters are then checked, until a point is reached where no reinforcement is required.

For slabs at least 200 mm thick, shear reinforcement in the form of links or bent-up bars may be provided. The area A_{sv} of this reinforcement is given by

$$A_{sv} \geqslant (v - v_c)ud/(0.87 f_{yv} \sin \alpha),$$ (12.6)

where f_{yv} is the characteristic strength of the shear reinforcement and α is the inclination of the shear reinforcement to the plane of the slab. In Equation 12.6, $(v - v_c)$ should not be taken as less than 0.4 N/mm². The shear reinforcement should be distributed evenly around the zone, along at least two critical perimeters, and the spacing should not exceed 1.5d.

A minimum amount of untensioned reinforcement adjacent to the upper concrete surface is recommended in TR 17, equal to 0.15% of the gross concrete cross-section, in order to enhance the shear resistance of the slab. This is additional to any reinforcement required for ultimate flexural strength.

An alternative to links or bent-up bars is to use prefabricated steel shear-heads. The design of these is not covered by BS8110, but a suitable method may be found in the American code ACI 318-77 (American Concrete Institute, 1977).

The vertical component of any tendons crossing the critical perimeter may be taken into account in assessing the shear resistance of a prestressed concrete flat slab. Only those tendons lying within the column lines should be considered as contributing to the shear resistance, and this is one reason why as many tendons as possible are usually placed in these locations. The maximum stress in the tendons should be taken as f_{pe}.

The total shear resistance, regardless of the type of reinforcement used, should not exceed twice that of the slab with no reinforcement.

Further information on the design of prestressed flat slabs may be found in Cope and Clark (1984) and Cement and Concrete Association (1980) and on their shear resistance in Regan (1985).

EXAMPLE 12.3 ■ ■

For the slab in Example 12.1, determine the shear resistance at an interior column.

The side of the critical perimeter is given by

$x_p = 500 + 2 \times 1.5 \times 270$

$= 1310 \, \text{mm}$.

The tendon profile shown in Fig. 12.10 was an approximate profile for initial estimation of the prestress force. A more practical profile, allowing a smooth change of curvature over the columns, is shown in Fig. 12.20. The inclination of the tendons at the first critical perimeter is approximately the same as that at the inflexion point of the tendon profile, i.e.

$$\theta \approx \tan^{-1}\left[\frac{4(e_1 + e_2)}{L}\right]$$

$$= \tan^{-1}(4 \times 140/7500) = 4.3°.$$

In the transverse direction, the width of the column line

$$= 500 + 2 \times 200/2 = 700 \, \text{mm}.$$

With a minimum tendon spacing of 75 mm, the maximum number of tendons within the column line is ten and, with tendon spacing of 330 mm between the column lines, this is also the total number of tendons crossing one side of the critical perimeter.

In the longitudinal direction, it is assumed that there are eight tendons lying within the column line, with similar inclination of the tendons at the first critical perimeter. The total upwards component of the tendon forces crossing the

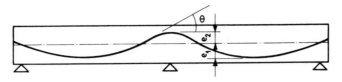

Fig. 12.20 Actual tendon profile.

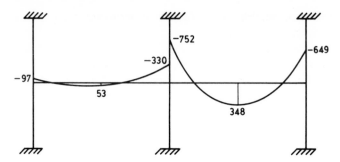

Fig. 12.21 Ultimate bending moment distribution for maximum shear force (kN m).

critical perimeter thus

$$= 2(10 + 8) \times 0.85 \times 185.9 \sin 4.3°$$
$$= 426.5 \, \text{kN.}$$

Since the ratio of imposed to dead load is greater than 1.25, and also the imposed load is greater than $5 \, \text{kN/m}^2$, the values of V_t and M_t for use in Equation 12.4 must be taken from an equivalent frame analysis with pattern loading. The bending moment diagram for this is shown in Fig. 12.21, with span AB loaded with 1.0 × dead load and span BC loaded with (1.4 × dead load + 1.6 × imposed load). By considering the equilibrium at the interior support, V_t is found to be 745.3 kN. Thus, the total effective shear force at the column is given by

$$V_{\text{eff}} = 745.3 \left[1 + \frac{1.5 \times (752 - 330)}{745.3 \times 1.31} \right] - 426.5$$

$$= 802.0 \, \text{kN.}$$

For the transverse direction,

$$A_s^* = (1770/460) \times 10 \times 150 = 5772 \, \text{mm}^2;$$

$$100 A_s^*/bd = (100 \times 5772)/(1310 \times 270) = 1.63.$$

Thus, from Table 7.2,

$$v_c = 0.80 \times (40/25)^{1/3} = 0.94 \, \text{N/mm}^2.$$

In the longitudinal direction,

$$A_s^* = (1770/460) \times 8 \times 150 = 4617 \, \text{mm}^2;$$

$$100 A_s^*/bd = (100 \times 4617)/(1310 \times 270) = 1.31;$$

and, from Table 7.2,

$$v_c = 0.74 \times (40/25)^{1/3} = 0.87 \, \text{N/mm}^2.$$

The total shear resistance along the critical perimeter thus

$$= 2 \times 1310(0.94 + 0.87) \times 270 \times 10^{-3}$$

$$= 1280.4 \, \text{kN} > V_{\text{eff}}.$$

■ ■

References

American Concrete Institute (1977) *Building Code Requirements for Reinforced Concrete*, ACI 318–77, Detroit.

Birkenmaier, M., Welbergen, G. H. and Winkler, N. (1986) *Post-Tensioned Concrete Flat Slabs*, BBR Ltd, Berne.

Cement and Concrete Association (1980) *FIP Recommendations for the Design of Flat Slabs in Post-Tensioned Concrete (Using Bonded and Unbonded Tendons)*, Slough.

Concrete Society (1979) *Flat Slabs in Post-Tensioned Concrete, With Particular Regard to the Use of Unbonded Tendons – Design Recommendations*, Technical Report No. 17, London.

Concrete Society (1984) *Post-Tensioned Flat-Slab Handbook*, Technical Report No. 25, London.

Cope, R.J. and Clark, L. A. (1984) *Concrete Slabs, Analysis and Design*, Elsevier Applied Science, London.

Regan, P. E. (1985) *The Punching Resistance of Prestressed Concrete Slabs*. Proceedings of the Institution of Civil Engineers, Part 2, Vol. 79.

Chapter 13

DESIGN EXAMPLES

Ref.	Example 13.1 Class 1 Member	13.1/1
	A highway bridge is required to span 24m, with main beams at 1·2m centres. The road surface is formed by a 180mm RC slab acting compositely with the beams. Loading $\qquad$ kN/m²	

A highway bridge is required to span 24m, with main beams at 1·2m centres. The road surface is formed by a 180mm RC slab acting compositely with the beams.

<u>Loading</u>   kN/m²

Slab	4·32
Finishes	2·40
Imposed load	10·30

<u>Material Properties</u>

Concrete : f_{cu} at 28 days = 52·5 N/mm² (beams)

f_{ci} at 7 days = 40·0 N/mm²

f_{cu} at 28 days = 30·0 N/mm² (slab)

Steel : f_{pu} = 1820 N/mm² (tendons)

f_y = 460 N/mm² (bars)

f_{yv} = 250 N/mm² (links)

<u>Allowable Stresses</u>

Table 3.3

Beams : f'_{max} = 20·0 N/mm²; f_{max} = 17·5 N/mm²

f'_{min} = −1·0 N/mm²; f_{min} = 0 N/mm²

Slab : f_{max} = 10·0 N/mm²

<u>Initial Estimate of Beam Depth</u>

For the beam spacing of 1·2m, choose a L/h ratio of 20.

$$\therefore h = \frac{24}{20} = 1·2m$$

Chosen section :

Ref.	Example 13.1　　Class 1 Member	13.1/2
	Beam section properties:	

Beam section properties:

$$w_i = 8.70 \, KN/m \qquad A_c = 3.694 \times 10^5 mm^2$$

$$Z_t = 80.75 \times 10^6 \, mm^3 \qquad I_b = 5.871 \times 10^{10} mm^4$$

$$Z_b = 119.32 \times 10^6 mm^3$$

<u>*Minimum Composite Section Size*</u>

$$w_d = 8.70 + 1.2(4.32 + 2.40) = 16.76 \, kN/m$$

$$w_s = 16.76 + 1.2 \times 10.30 \qquad = 29.12 \, kN/m$$

$$M_i = 8.70 \times \frac{24^2}{8} = 626.4 \, kNm$$

$$M_d = 16.76 \times \frac{24^2}{8} = 1206.7 \, kNm$$

$$M_s = 29.12 \times \frac{24^2}{8} = 2096.6 \, kNm$$

Initially assume that $\alpha = 0.90$, $\beta = 0.75$

Eq. 10.5

$$Z_{b,comp} \geqslant \frac{0.90(2096.6 - 1206.7) \times 10^6}{(0.75 \times 20 - 0) + \frac{1}{119.32}(0.75 \times 626.4 - 0.90 \times 1206.7)}$$

$$= 81.43 \times 10^6 \, mm^3$$

For the composite section:

$$\left(\frac{1200 \times 180^2}{2} + 3.694 \times 10^5 \times 908\right) = \bar{y}\left(1200 \times 180 + \frac{3.694}{\times 10^5}\right)$$

$$\therefore \bar{y} = 606 \, mm$$

$$I_{comp} = \frac{1200 \times 180^3}{12} + 1200 \times 180(606 - 90)^2$$

$$+ \, 5.871 \times 10^{10} + 3.694 \times 10^5 (908 - 606)^2$$

$$= 1.505 \times 10^{11} mm^4$$

$$Z_{b,comp} = \frac{1.505 \times 10^{11}}{(1400 - 606)} = 189.54 \times 10^6 mm^3$$

Section size adequate

<u>*Prestress Force*</u>

Assume cover to the ducts = 50mm and duct dia. = 65mm

Ref.	*Example 13.1 Class 1 Member*	13.1/3
Eq. 10.6(a)	$e_{max} = 1400 - 908 - (50 + \frac{65}{2}) = 410 \, mm$	

$$P_i \geqslant \frac{80.75 \times 10^6(-1.0) - 626.4 \times 10^6}{0.90\left(\frac{80.75 \times 10^6}{3.694 \times 10^5} - 410\right)}$$

i.e. $P_i \leqslant 4105.1 \, kN$

Eq. 10.6(b) Also, $P_i \geqslant \dfrac{119.32 \times 10^6\left[\dfrac{2096.6}{189.54} + 1206.7\left(\dfrac{1}{119.32} - \dfrac{1}{189.54}\right)\right]}{0.75\left(\dfrac{119.32 \times 10^6}{3.694 \times 10^5} + 410\right)}$

i.e. $P_i \geqslant 3214.0 \, kN$

Table 2.2 For 4 no. 4/15.2 mm drawn strands,

$$P_i = 4 \times 4 \times 165 \times 0.7 \times 1820 \times 10^{-3}$$

$$= 3363.4 \, kN$$

$$e_{max} = 410 - \frac{130}{4} = 377 \, mm$$

Preliminary Check on Stresses

Transfer:

$$f_t = \frac{0.90 \times 3363.4 \times 10^3}{3.694 \times 10^5} - \frac{0.90 \times 3363.4 \times 10^3 \times 377}{80.75 \times 10^6}$$

$$+ \frac{626.4 \times 10^6}{80.75 \times 10^6}$$

$$= 8.19 - 14.13 + 7.76 = 1.82 \, N/mm^2 \, (> f'_{min})$$

$$f_b = 8.19 + 14.13 \times \frac{80.75}{119.32} - \frac{626.4 \times 10^6}{119.32 \times 10^6}$$

$$= 8.19 + 9.56 - 5.25 = 12.50 \, N/mm^2 (< f'_{max})$$

Service:

$$f_t = 8.19 \times \frac{0.75}{0.90} - 14.13 \times \frac{0.75}{0.90} + \frac{1206.7 \times 10^6}{80.75 \times 10^6}$$

$$+ \frac{(2096.6 - 1206.7) \times 10^6 \times (606 - 180)}{1.505 \times 10^{11}}$$

$$= 6.83 - 11.78 + 14.94 + 2.52$$

$$= 12.51 \, N/mm^2 \, (< f_{max})$$

Ref.	Example 13.1 Class 1 Member	13.1/4

$$f_b = 6.83 + 9.56 \times \frac{0.75}{0.90} - \frac{1206.7 \times 10^6}{119.32 \times 10^6}$$

$$- \frac{889.9 \times 10^6}{189.54 \times 10^6}$$

$$= 6.83 + 7.97 - 10.10 - 4.70 = 0 \ N/mm^2 \ (= f_{min})$$

$$f_{t_{slab}} = \frac{889.9 \times 10^6 \times 606}{1.505 \times 10^{11}} = 3.58 \ N/mm^2 \ (< f_{max})$$

Estimate of Losses

(i) Friction

Assume $\mu = 0.25$; $K = 17 \times 10^{-4}/m$

Assumed profile :

12m   12m   377

$$\frac{1}{r_{ps}} = \frac{8 \times 0.377}{24^2} \qquad \therefore r_{ps} = 190.98 \ m$$

$$p = 3363.4 \left[1 - e^{-\left(\frac{0.25}{190.98} + 17 \times 10^{-4} \right)} \right] = 10.1 \ kN/m$$

For 5mm anchorage draw-in,

Eq.4.11 $\quad x = \left(\dfrac{5 \times 195 \times 10^3 \times 2640}{10.1} \right)^{\frac{1}{2}} \times 10^{-3} = 15.96 \ m$

Eq.4.10 $\quad \therefore \Delta P_A = 2 \times 10.1 \times 15.96 = 322.4 \ kN$

Prestress force after friction losses :

Tensioning end: $P = 3363.4 - 322.4 = 3041.0 \ kN$

Midspan : $P = 3363.4 - 322.4 + 12 \times 10.1 = 3162.2 \ kN$

Dead end : $P = 3363.4 - 24 \times 10.1 = 3121.0 \ kN$

(ii) Elastic Shortening

Average force after friction losses = 3108.1 kN

$$\therefore f_{pi} = \frac{3108.1 \times 10^3}{2640} = 1177 \ N/mm^2$$

$$m = \frac{195}{28} = 7.0 \ ; \quad r = \left(\frac{5.871 \times 10^{10}}{3.694 \times 10^5} \right)^{\frac{1}{2}} = 399 \ mm$$

Ref.	Example 13.1 Class 1 Member	13.1/5

At midspan,

$$f_{co} = \frac{1177}{\left[7.0 + \frac{3.694 \times 10^5}{2640\left(1 + \frac{377^2}{399^2}\right)}\right]} - \frac{626.4 \times 10^6 \times 377}{5.871 \times 10^{10}}$$

$$= 14.54 - 4.02 = 10.52 \ N/mm^2$$

At the supports,

$$f_{co} = \frac{1177}{\left[7.0 + \frac{3.694 \times 10^5}{2640}\right]} = 8.01 \ N/mm^2$$

$\therefore$ Average $f_{co} = \frac{1}{2}(10.52 + 8.01) = 9.27 \ N/mm^2$

$\therefore \Delta f_p = \frac{1}{2} \times 7.0 \times 9.27 = 32.5 \ N/mm^2$

$\qquad\qquad\qquad (2.5\% \ of \ 0.7f_{pu})$

$\alpha = 0.91$

(ii) Creep

$$\varepsilon_c = \frac{1.8}{28 \times 10^3} \times 9.27 = 5.96 \times 10^{-4}$$

$\therefore \Delta f_p = 5.96 \times 10^{-4} \times 195 \times 10^3 = 116.2 \ N/mm^2$

(iii) Shrinkage

For outdoor exposure,

$$\Delta f_p = 100 \times 10^{-6} \times 195 \times 10^3 = 19.5 \ N/mm^2$$

(v) Relaxation

For Class 2 strands,

$$\Delta f_p = 1.5 \times 0.025 \times 1177 = 44.1 \ N/mm^2$$

$\therefore$ Total long-term losses = $179.8 \ N/mm^2$

$\qquad\qquad\qquad (14.1\% \ of \ 0.7f_{pu})$

$\beta = 0.77$

<u>Prestress Force Distribution</u>

m	0	3	6	9	12	15	18	21	24
αP_i	2956.9	2987.2	3017.5	3047.8	3078.1	3108.4	3097.5	3067.2	3036.9
βP_i	2482.7	2513.0	2543.3	2573.6	2603.9	2634.2	2623.3	2593.0	2562.7

Ref.	Example 13.1 Class 1 Member	13.1/6

Eq.'s

Cable Zone

10.7(b)
$$e \geqslant \frac{1}{\beta P_i}\left(0.37 M_d + 0.63 M_s\right) - 323.0$$

10.7(a)
$$e \leqslant \frac{1}{\alpha P_i}\left(M_i + 80.75 \times 10^6\right) + 218.6$$

m	0	3	6	9	12	15	18	21	24
M_i	0	274.1	469.8	587.3	626.4	587.3	469.8	274.1	0
M_d	0	527.9	905.1	1131.3	1206.7	1131.3	905.1	527.9	0
M_s	0	917.3	1572.5	1965.6	2096.6	1965.6	1572.5	917.3	0
$e \geqslant$	-323	-15	198	321	356	306	182	-25	-323
$e \leqslant$	246	337	401	438	448	434	396	334	245

The maximum midspan eccentricity of 377mm is thus within the cable zone.

Tendon Profiles

Tendon \ m	0	3	6	9	12
1	-442	-126	100	235	280
2	-242	43	247	369	410
3 & 4	342	371	393	406	410
Resultant	0	165	283	354	377

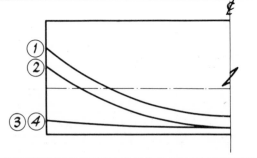

Ref.	Example 13.1 Class 1 Member	13.1/7

Revised Prestress Force Distribution

Tendon	r_{ps}	p	x	ΔP_A
1	99.72	3.5	13.56	94.9
2	110.43	3.3	13.96	92.1
3 & 4	1058.82	1.6	20.05	64.2

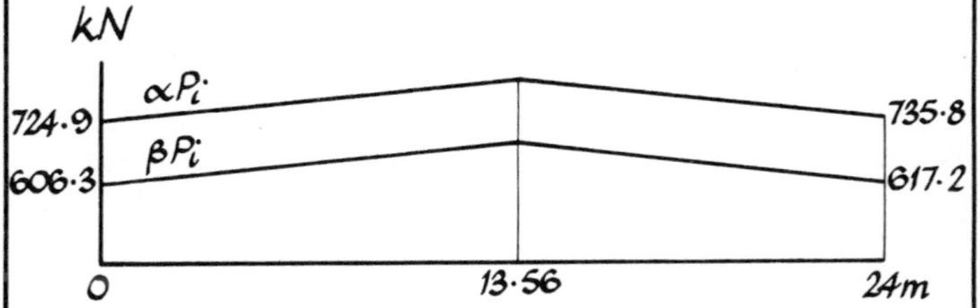

kN

(1)

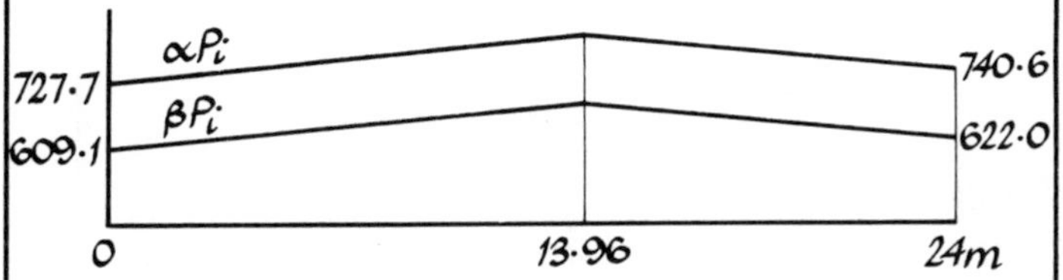

(2)

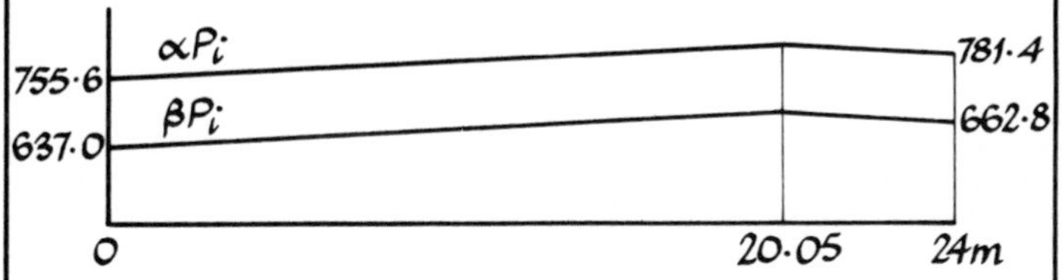

(3) & (4)

Total prestress force:

m	0	3	6	9	12	15	18	21	24
αP_i	2963.8	2993.8	3023.8	3053.8	3083.8	3096.8	3086.0	3069.2	3039.2
βP_i	2489.4	2519.4	2549.4	2579.4	2609.4	2622.4	2611.6	2594.8	2564.8

Ref.	Example 13.1 Class 1 Member								13.1/8

Revised cable zone:

m	0	3	6	9	12	15	18	21	24
$e \geqslant$	-323	-16	197	319	354	309	185	-25	-323
$e \leqslant$	246	337	401	437	448	434	397	334	245

Ultimate Strength

Ult. uniform load $= 1.4 \times 16.76 + 1.6 \times 1.2 \times 10.3$

$\qquad = 43.24 \text{ kN/m}$

Max. ult. BM $= \dfrac{43.24 \times 24^2}{8} = 3113.3 \text{ kNm}$

Assume that the tendons have yielded and that the concrete stress block is within the beam upper flange.

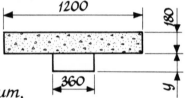

For equilibrium,

$0.87 \times 1820 \times 2640 = 0.45 \times 30 \times 180 \times 1200$

$\qquad\qquad\qquad + 0.45 \times 52.5 \times 360 y$

$\qquad\qquad \therefore y = 149 \text{ mm}$

The depth of the upper flange is 145 mm, so that the above assumption is approximately correct.

Take $0.9x = 325 \text{ mm}$

$\varepsilon_{pb} = \dfrac{2609.4 \times 10^3}{2640 \times 195 \times 10^3} + \dfrac{(1285 - 361) \times 0.0035}{361}$

$\qquad = 0.0140 \ (> \varepsilon_2 = 0.0131)$

$\therefore M_u = \left[0.45 \times 30 \times 1200 \times 180 \left(1285 - \dfrac{180}{2}\right) \right.$

$\qquad \left. + 0.45 \times 52.5 \times 360 \times 145 \left(1285 - 180 - \dfrac{145}{2}\right) \right] \times 10^{-6}$

$\qquad = 4757.9 \text{ kNm}$

Ultimate strength adequate

Ref.	Example 13.1 Class 1 Member	13.1/9
	Horizontal Shear At midspan, the total force in the slab $= 0.45 \times 30 \times 1200 \times 180 \times 10^{-3}$ $= 2916.0 \ kN$ $\therefore V_{h_{avge}} = \dfrac{2916.0 \times 10^3}{12 \times 10^3 \times 360} = 0.68 \ N/mm^2$ $\therefore V_{h_{max}} = 2 \times 0.68 = 1.36 \ N/mm^2$	
Table 10.2	Allowable $V_{h_{max}} = 1.8 \ N/mm^2$ $\therefore$ Nominal links required i.e. $A_h = \dfrac{0.15}{100} \times 360 \times 10^3 = 540 \ mm^2/m$	Use R10-275 (571mm²/m)
	Vertical Shear (i) *Uncracked sections* The principal tensile stresses will be checked at the composite section centroid. (a) Dead load $f_{cp} = \dfrac{\beta P_i}{A_c} - \dfrac{\beta P_i \ e \ (908 - \bar{y})}{I_b}$ $f_s = \dfrac{Nett \ V_d}{0.67 \times 150 \times 1220}$	

m	0	3	6	9	
f_{cp}	6.60	4.58	3.13	2.24	
V_d	281.6	211.2	140.8	70.4	
$\sum \beta P_i \sin\theta$	135.1	103.4	70.2	35.8	Tendons 1+2
$Nett \ V_d$	146.5	107.8	70.6	34.6	
f_s	1.19	0.88	0.58	0.28	
f_{prt}	0.21	0.16	0.10	0.03	

Eq. 7.2

Ref.	Example 13.1 Class 1 Member	13.1/10

(b) Service load

$$\Sigma f_s = \frac{\text{Nett } V_d}{0.67 \times 150 \times 1220} + \frac{(V_s - V_d)}{0.67 \times 150 \times 1400}$$

m	0	3	6	9
$V_s - V_d$	237.3	178.0	118.7	59.3
Σf_s	2.88	2.15	1.42	0.70
f_{prt}	1.08	0.85	0.55	0.20

Eq.7.2

Eq.7.4

Allowable $f_{prt} = 0.24 \times (52.5)^{1/2} = 1.74 \, N/mm^2$

(ii) Cracked sections

(a) Dead load

$$f_{pt} = \frac{\beta P_i}{A_c} + \frac{\beta P_i \, e \times 492}{I_b}$$

$$M_o = \frac{0.8 f_{pt} I_b}{492} \; ; \; \frac{f_{pe}}{f_{pu}} = \frac{2609.4 \times 10^3}{2640 \times 1820} = 0.54$$

m	3	6	9	12
f_{pt}	10.09	12.70	14.35	15.02
M_d	739.1	1267.0	1583.8	1689.4
V_d	211.2	140.8	70.4	0
M_o	963.2	1212.4	1369.9	1433.9
d	893	1011	1082	1105
v_c	0.79	0.76	0.74	0.73
V_{cr}	350.2	216.4	146.0	85.7

Eq.7.13

(b) Service load

Eq.10.2

$$M_o = M_d + \left(0.8 f_{pt} - \frac{M_d}{119.32 \times 10^6}\right) \times 189.54 \times 10^6,$$

with $\gamma_m = 1.0$ for M_d

Uncracked shear resistance adequate

Ref.	Example 13.1 Class 1 Member	13.1/11

m	3	6	9	12
M_o	1219·7	1394·0	1510·4	1568·7
V_s	389·2	259·4	129·7	0
M_s	1362·1	2335·0	2918·7	3113·3
d	1073	1191	1262	1285
v_c	0·74	0·72	0·70	0·70
V_{cr}	432·9	246·0	161·0	95·6

Eq. 7·13

Nominal shear reinforcement is required for all but the central 3 m.

Nominal links :

$$\frac{A_{sv}}{s_v} = \frac{0·4 \times 150}{0·87 \times 250} = 0·276$$

Use R10-500
$\left(\dfrac{A_{sv}}{s_v} = 0·314\right)$

Ref.	Example 13.1 Class 1 Member	13.1/12

<u>End Blocks</u>

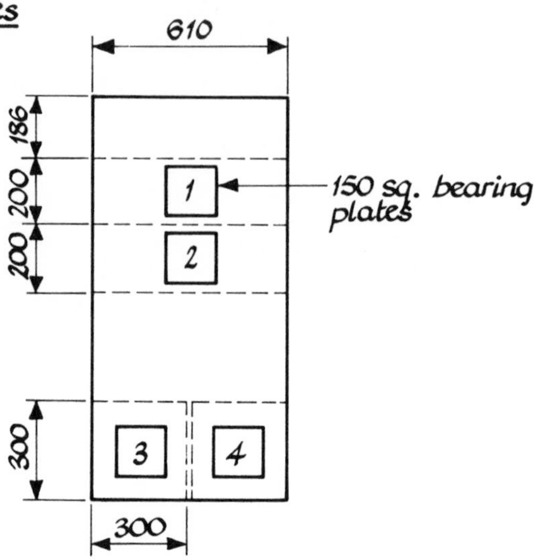

For tendons 1 & 2:

Horiz. side : $\dfrac{y_{po}}{y_o} = \dfrac{75}{305} = 0.25$

Vert. side : $\dfrac{y_{po}}{y_o} = \dfrac{75}{100} = 0.75$

$P_i = 4 \times 165 \times 0.70 \times 1820 \times 10^{-3} = 840.8$ kN

Max. $F_{bst} = 0.23 \times 840.8 = 193.4$ kN

A_{sv} req'd. $= \dfrac{193.4 \times 10^3}{200} = 967$ mm^2

Extent of links: 61 – 610 mm

Use 5R12
(A_{sv} = 1130)

For tendons 3 & 4:

Horiz.& vert.: $\dfrac{y_{po}}{y_o} = \dfrac{75}{150} = 0.50$
sides

Max. $F_{bst} = 0.17 \times 840.8 = 142.9$ kN

A_{sv} req'd. $= \dfrac{142.9 \times 10^3}{200} = 715$ mm^2

Extent of links: 30 – 300 mm

Use 4R12
(A_{sv} = 904)

Ref	Example 13.1 Class 1 Member	13.1/13
	For tendons 1+2 as a combined anchorage the equivalent prism is 610 × 772 mm. Equivalent loaded area has side $(2 \times 150^2)^{1/2}$ or 212 mm Horiz. side: $\dfrac{y_{po}}{y_o} = \dfrac{106}{305} = 0.35$ Vert. side: $\dfrac{y_{po}}{y_o} = \dfrac{106}{386} = 0.27$ Max. $F_{bst} = 2 \times 840.8 \times 0.23 = 386.8$ kN A_{sv} req'd. $= \dfrac{386.8 \times 10^3}{200} = 1934$ mm^2 Extent of links: 77–770 mm For tendons 3+4, the equivalent prism is 610 × 300 mm Horiz. side: $\dfrac{y_{po}}{y_o} = \dfrac{106}{305} = 0.35$ Vert. side: $\dfrac{y_{po}}{y_o} = \dfrac{106}{150} = 0.71$ Max. $F_{bst} = 2 \times 840.8 \times 0.22 = 370.0$ kN A_{sv} req'd. $= \dfrac{370.0 \times 10^3}{200} = 1850$ mm^2 Extent of links: 61–610 mm	 Use 10R12 $(A_{sv} = 2260)$ Use 9R12 $(A_{sv} = 2034)$

Ref.	Example 13.1 Class 1 Member	13.1/14

<u>Deflections</u>

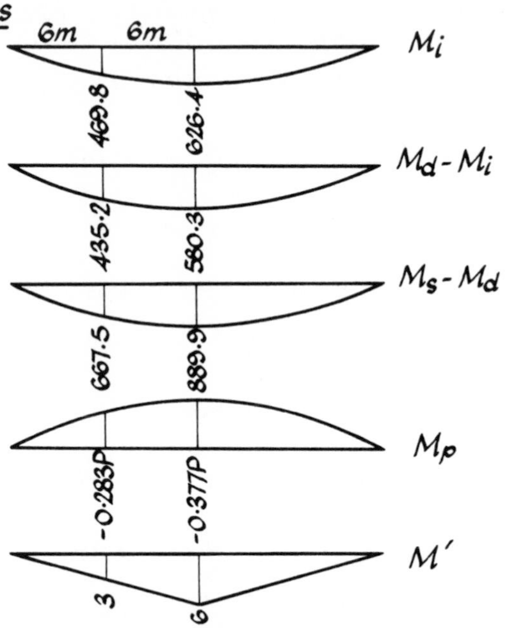

6m   6m M_i

469·8   626·4

$M_d - M_i$

435·2   580·3

$M_s - M_d$

667·5   889·9

M_p

-0·283P   -0·377P

M'

3   6

(i) *Short-term deflections* :

$$\delta_i = \frac{2 \times 12}{6E_c I_b} \left[\begin{array}{l} 4(469 \cdot 8 - 0 \cdot 283P)(3) \\ + (626 \cdot 4 - 0 \cdot 377P)(6) \end{array} \right]$$

For average $\alpha P_i = 3045 \cdot 6$ kN,

$$\delta_i = -\frac{31344}{E_c I_b}$$

Table
2.1

For $f_{ci} = 40$ N/mm², $E_c = 28$ kN/mm²

$$\therefore \delta_i = \frac{-31344}{28 \times 10^6 \times 5 \cdot 871 \times 10^{-2}} = -0 \cdot 0191 \text{ m}$$

$$\delta_i = -\frac{L}{1259}$$

$$\delta_d - \delta_i = \frac{2 \times 12}{6E_c I_b} \left[4(435 \cdot 2)(3) + (580 \cdot 3)(6) \right]$$

$$= \frac{34817}{28 \times 10^6 \times 5 \cdot 871 \times 10^{-2}} = 0 \cdot 0212 \text{ m}$$

$$\therefore \delta_d = -0 \cdot 0191 + 0 \cdot 0212$$

$$= 0 \cdot 0021 \text{ m}$$

$$\delta_d = \frac{L}{1143}$$

Ref.	Example 13.1 Class 1 Member	13.1/15
Tables 2·1 & 6·3 Fig.2·6	*(ii) Long-term deflections* After 1 year, $E_c = 32 \ kN/mm^2$ $t_{eff} = \dfrac{2 \times 3 \cdot 694 \times 10^5}{3326} = 222 \ mm$ $\therefore \phi = 0 \cdot 8$ and $E_{c_{eff}} = \dfrac{32}{(1 + 0 \cdot 8)}$ $\qquad\qquad\qquad = 17 \cdot 8 \ kN/mm^2$ Short-term deflections under non-permanent load: $\delta_s - \delta_d = \dfrac{2 \times 12}{6 E_c I_{comp}} \left[4(667 \cdot 5)(3) + (889 \cdot 9)(6) \right]$ $\qquad = \dfrac{53398}{32 \times 10^6 \times 1 \cdot 505 \times 10^{-1}} = 0 \cdot 0111 \ m$ Long-term deflections under permanent load: For average $\beta P_i = 2571 \cdot 2 \ kN$, $\delta_i = \dfrac{-20607}{17 \cdot 8 \times 10^6 \times 5 \cdot 871 \times 10^{-2}} = -0 \cdot 0197 \ m$ $\delta_d - \delta_i = 0 \cdot 0212 \times \dfrac{28}{17 \cdot 8} = 0 \cdot 0331 \ m$ $\therefore \delta_t = -0 \cdot 0197 + 0 \cdot 0331 + 0 \cdot 0111$ $\qquad = 0 \cdot 0245 \ m$	 $\delta_t = \dfrac{L}{980}$

Ref.	*Example 13.2 Class 2 Member*	13.2/1

Pretensioned double-T roof beams are required to span 15 m

<u>*Loading*</u> kN/m^2

    Finishes   2·0

    Imposed Load 0·75

<u>*Material Properties*</u>

 Concrete : f_{cu} *at 28 days* = 50 N/mm^2

      f_{ci} *at 7 days* = 35 N/mm^2

 Steel : f_{pu} = 1860 N/mm^2

<u>*Allowable Stresses*</u>

 $f'_{max.}$ = 17·5 N/mm^2 $f_{max.}$ = 16·7 N/mm^2

 $f'_{min.}$ = −2·7 N/mm^2 $f_{min.}$ = −3·2 N/mm^2

<u>*Initial Estimate of Beam Depth*</u>

Assuming a L/h ratio of 30, the depth of beam required is given by :

$$h = \frac{15}{30} = 0·5m$$

Chosen section :

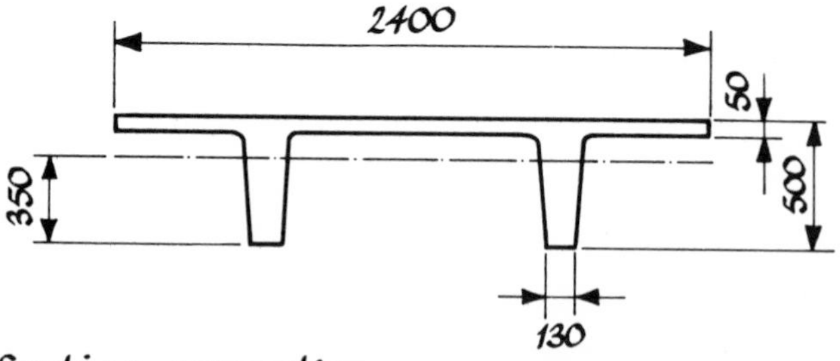

Section properties :

W_i = 6·36 kN/m A_c = 2·648 × 10^5 mm^2

Z_t = 40·23 × $10^6 mm^3$ I = 6·041 × 10^9 mm^4

Z_b = 17·26 × $10^6 mm^3$

Ref.	Example 13.2 Class 2 Member	13.2/2
	$w_S = 6 \cdot 36 + 2 \cdot 4 (2 \cdot 0 + 0 \cdot 75)$ $\quad = 12 \cdot 96 \quad N/m$ $M_i = \dfrac{6 \cdot 36 \times 15^2}{8} = 178 \cdot 9 \ kNm$ $M_S = \dfrac{12 \cdot 96 \times 15^2}{8} = 364 \cdot 5 \ kNm$ <u>Minimum Section Size</u> Initially assume that $\alpha = 0 \cdot 90$, $\beta = 0 \cdot 75$	
Eq. 9.3(a)	$Z_t \geqslant \dfrac{(0 \cdot 90 \times 364 \cdot 5 - 0 \cdot 75 \times 178 \cdot 9) \times 10^6}{0 \cdot 90 \times 16 \cdot 7 - 0 \cdot 75 (-2 \cdot 7)}$ $\quad = 11 \cdot 37 \times 10^6 \ mm^3$	
Eq. 9.3(b)	$Z_b \geqslant \dfrac{(0 \cdot 90 \times 364 \cdot 5 - 0 \cdot 75 \times 178 \cdot 9) \times 10^6}{0 \cdot 75 \times 17 \cdot 5 - 0 \cdot 90 (-3 \cdot 2)}$ $\quad = 12 \cdot 11 \times 10^6 \ mm^3$	Section size adequate
	<u>Magnel Diagram</u> Assuming a total of 12 no. strands, the max. eccentricity is taken as: $e_{max.} = 350 - 100$ $\qquad = 250 \ mm$	
9.5(a)	$10^8 \times \dfrac{1}{P_i} \geqslant 0 \cdot 313 e - 47 \cdot 5 \quad (for\ e \geqslant 152\ mm)$	
9.5(b)	$10^8 \times \dfrac{1}{P_i} \geqslant 0 \cdot 187 e + 12 \cdot 2$	
9.5(c)	$10^8 \times \dfrac{1}{P_i} \geqslant -0 \cdot 244 e + 37 \cdot 0 \ (for\ e \leqslant 152\ mm)$	
9.5(d)	$10^8 \times \dfrac{1}{P_i} \leqslant 0 \cdot 243 e + 15 \cdot 8$	

Ref.	Example 13.2　　Class 2　Member 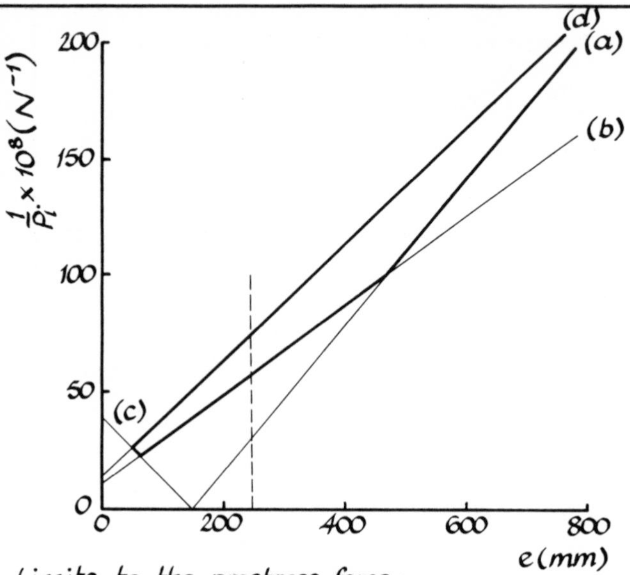	13.2/3

Limits to the prestress force:

$$P_i \geqslant 1308.3 \ kN$$

$$P_i \leqslant 1695.5 \ kN$$

For 12 no. 12.9 mm dia. super strands,

$$P_i = 12 \times 100 \times 0.7 \times 1860 \times 10^{-3}$$

$$= 1562.4 \ kN$$

Estimate of Losses

(i) Elastic shortening

$$f_{pi} = 0.7 \times 1860 = 1302 \ N/mm^2$$

$$m = \frac{195}{27} = 7.2$$

$$r = \left(\frac{6.041 \times 10^9}{2.648 \times 10^5} \right)^{\frac{1}{2}} = 151 \ mm$$

At midspan,

Eq.4.4

$$f_{co} = \left[\frac{1302}{7.2 + \dfrac{2.648 \times 10^5}{1200 \left(1 + \frac{250^2}{151^2} \right)}} \right] = 19.67 \ N/mm^2$$

Ref.	Example 13.2 Class 2 Member	13.2/4
	At the supports, assume that the 4no. upper strands are de-bonded. $\therefore f_{co} = \left[\dfrac{1302}{7 \cdot 2 + \dfrac{2 \cdot 648 \times 10^5}{800\left(1 + \dfrac{250^2}{151^2}\right)}} \right] = 13 \cdot 61 \; N/mm^2$ $\therefore$ Average $f_{co} = \frac{1}{2}(19 \cdot 67 + 13 \cdot 61) = 16 \cdot 64 \; N/mm^2$ $\therefore \Delta f_p = 7 \cdot 2 \times 16 \cdot 64 = 119 \cdot 8 \; N/mm^2$ (9.2% loss) **(ii) Creep** Concrete stress at level of tendons at midspan $= 19 \cdot 67 - \dfrac{178 \cdot 9 \times 10^6 \times 250}{6 \cdot 041 \times 10^9} = 12 \cdot 27 \; N/mm^2$ $\therefore$ Average stress $= \frac{1}{2}(13 \cdot 61 + 12 \cdot 27) = 12 \cdot 94 \; N/mm^2$ $\therefore \varepsilon_c = \dfrac{1 \cdot 8}{27 \times 10^3} \times 12 \cdot 94 = 8 \cdot 63 \times 10^{-4}$ $\therefore \Delta f_p = 8 \cdot 63 \times 10^{-4} \times 195 \times 10^3 = 168 \cdot 2 \; N/mm^2$ **(iii) Shrinkage** For indoor exposure, $\varepsilon_{sh} = 300 \times 10^{-6}$ $\therefore \Delta f_p = 300 \times 10^{-6} \times 195 \times 10^3 = 58 \cdot 5 \; N/mm^2$ **(iv) Relaxation** For Class 2 strands, $\Delta f_p = 1 \cdot 2 \times 0 \cdot 025 \times 1302 = 39 \cdot 1 \; N/mm^2$ $\therefore$ Total long-term loss $= 265 \cdot 8 \; N/mm^2$ $(20 \cdot 4\%$ loss$)$ <u>Cable Zone</u>	$\alpha = 0 \cdot 91$ $\beta = 0 \cdot 70$
Eq's 9.6(a) 9.6(b) 9.6(c) 9.6(d)	Limits are given by: $e \leqslant 228 \cdot 3 + 7 \cdot 033 \times 10^{-7} M_i$ $e \leqslant 147 \cdot 3 + 7 \cdot 033 \times 10^{-7} M_i$ $e \geqslant -462 \cdot 4 + 9 \cdot 143 \times 10^{-7} M_s$ $e \geqslant -115 \cdot 7 + 9 \cdot 143 \times 10^{-7} M_s$	

Ref.	Example 13·2 Class 2 Member	13. 2/5

m	0	2·5	5·0	7·5
M_i	0	99·4	159·0	178·9
M_s	0	202·5	324·0	364·5
$e \geqslant$	-116	69	181	218
$e \leqslant$	147	217	259	273

For straight tendon profiles, the tendons lie outside the cable zone near the supports. The upper 4no. tendons will thus be de-bonded in these regions.

For the point where the resultant prestress force just lies within the cable zone,

$$147·3 + 7·033 \times 10^{-7} M_i = 250$$

$$\therefore M_i = 146·0 \ kNm$$

$$\therefore \frac{6·36}{2} x (15-x) = 146·0$$

$$\therefore x = 4·29 m$$

For a reduced prestress force of $\frac{2}{3} \times 1562·4$
$$= 1041·6 \ kN,$$

$$e \geqslant -140·9 + 1·372 \times 10^{-6} M_s$$

$$e \leqslant 253·5 + 1·055 \times 10^{-6} M_i$$

Eq.8·1   Transmission length $= \dfrac{240 \times 12·9}{(35)^{1/2}}$

$$= 523 \ mm$$

Ref.	*Example 13.2 Class 2 Member*	*13.2/6*

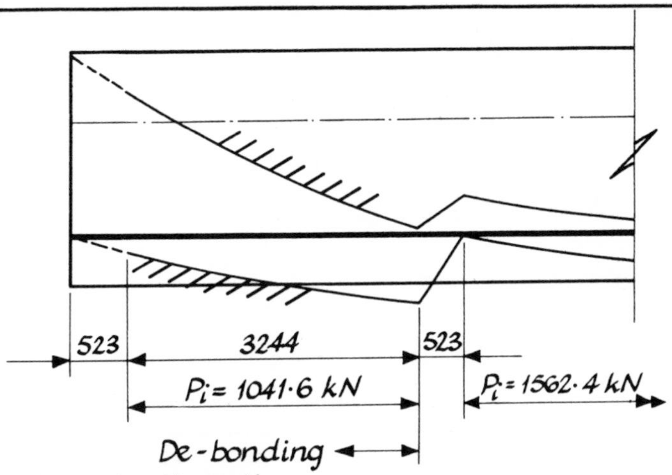

523   3244   523

$P_i = 1041 \cdot 6\ kN$ $P_i = 1562 \cdot 4\ kN$

De - bonding

<u>Ultimate Strength</u>

Ult. uniform load $= 1 \cdot 4 \times 11 \cdot 16 + 1 \cdot 6 \times 0.75 \times 2 \cdot 4$

$= 18 \cdot 50\ kN/m$

$\therefore$ Max. ult. BM $= \dfrac{18 \cdot 50 \times 15^2}{8} = 520 \cdot 3\ kNm$

Assuming that the tendons have yielded and that $0 \cdot 9x \leq 50$,

$0 \cdot 87 \times 1860 \times 1200 = 0 \cdot 9x \times 2400 \times 0 \cdot 45 \times 50$

$\therefore x = 40\ mm$ (i.e. $0 \cdot 9x < 50$)

$\varepsilon_{pb} = \dfrac{0 \cdot 70 \times 1562 \cdot 4 \times 10^3}{1200 \times 195 \times 10^3} + \dfrac{(400-40) \times 0 \cdot 0035}{40}$

$= 0 \cdot 0362$ $(> \varepsilon_2 = 0 \cdot 0133)$

$\therefore M_u = 0 \cdot 87 \times 1860 \times 1200 (400 - 0 \cdot 45 \times 40) \times 10^{-6}$

$= 741 \cdot 8\ kNm$ **Ultimate strength adequate**

<u>Shear Resistance</u>

(a) Uncracked sections

Since the neutral axis of the section lies within the web, Eq. 7·6 may be used.

Ref.	Example 13.2 Class 2 Member	13.2/7

At the critical section for shear,

$$f_{cp} = \frac{0.70 \times 1041.6 \times 10^3}{2.648 \times 10^5} = 2.75 \ N/mm^2$$

Table 7.1

$$\therefore V_{co} = 1.73 \times 2 \times 130 \times 500 \times 10^{-3} = 224.9 \ kN$$

$$V = 18.50 \times (7.5 - 0.45) = 130.4 \ kN$$

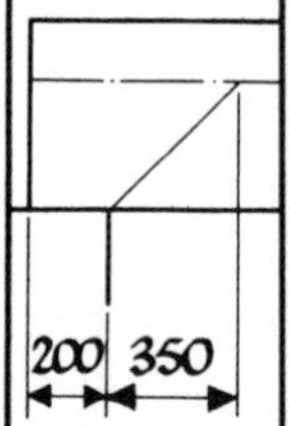

(b) Cracked sections

Eq.7.13

m	2.00	4.00	4.29	5.00	7.50
M_o	183.8	183.8	183.8	275.8	275.8
M	240.5	407.0	425.0	462.5	520.4
V	101.8	64.8	59.4	46.3	0
V_c	0.57	0.57	0.57	0.66	0.66
V_{cr}	121.1	72.6	69.0	77.7	50.1

Deflections

200 | 350

$L_t = 523$

Uncracked shear resistance adequate

Cracked shear resistance adequate

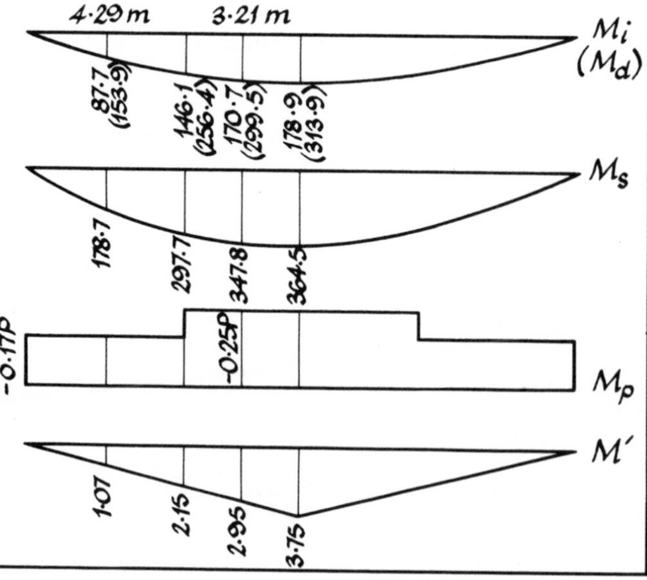

Ref.	Example 13.2 Class 2 Member	13.2/8
	(a) *Initial camber*	

$$\delta_i = \frac{2 \times 4 \cdot 29}{6E_c I} \left[4(87 \cdot 7 - 0 \cdot 17P)(1 \cdot 07) + (146 \cdot 1 - 0 \cdot 17P)(2 \cdot 15) \right]$$

$$+ \frac{2 \times 3 \cdot 21}{6E_c I} \left[\begin{array}{c} (146 \cdot 1 - 0 \cdot 25P)(2 \cdot 15) + 4(170 \cdot 7 - 0 \cdot 25P)(2 \cdot 95) \\ + (178 \cdot 9 - 0 \cdot 25P)(3 \cdot 75) \end{array} \right]$$

For $P = \alpha P_i = 0 \cdot 91 \times 1562 \cdot 4 = 1421 \cdot 8 \, kN,$

$$\delta_i = \frac{-4759 \cdot 2}{27 \times 10^6 \times 6 \cdot 041 \times 10^{-3}} = -0 \cdot 0292 \, m \qquad \delta_i = -\frac{L}{514}$$

(b) *Long-term deflection*

Long-term deflection under permanent load:

$$\delta_d = \frac{2 \times 4 \cdot 29}{6E_{c_{eff}} I} \left[4(153 \cdot 9 - 0 \cdot 17P)(1 \cdot 07) + (256 \cdot 4 - 0 \cdot 17P)(2 \cdot 15) \right]$$

$$+ \frac{2 \times 3 \cdot 21}{6E_{c_{eff}} I} \left[\begin{array}{c} (256 \cdot 4 - 0 \cdot 25P)(2 \cdot 15) + 4(299 \cdot 5 - 0 \cdot 25P)(2 \cdot 95) \\ + (313 \cdot 9 - 0 \cdot 25P)(3 \cdot 75) \end{array} \right]$$

For $P = \beta P_i = 0 \cdot 70 \times 1562 \cdot 4 = 1093 \cdot 7 \, kN,$

$$\delta_d = \frac{473 \cdot 1}{E_{c_{eff}} I}$$

$t_{eff} = 126 \, mm \, ; \, \phi = 1 \cdot 5$

$$\therefore E_{c_{eff}} = \frac{32}{(1 + 1 \cdot 5)} = 12 \cdot 8 \, kN/mm^2$$

$$\therefore \delta_d = \frac{473 \cdot 1}{12 \cdot 8 \times 10^6 \times 6 \cdot 041 \times 10^{-3}} = 0 \cdot 0061 \, m$$

Short-term deflection under non-permanent load:

$$\delta_s - \delta_d = \frac{2 \times 4 \cdot 29}{6E_c I} \left[4(178 \cdot 7 - 153 \cdot 9)(1 \cdot 07) + (297 \cdot 7 - 256 \cdot 4)(2 \cdot 15) \right]$$

$$+ \frac{2 \times 3 \cdot 21}{6E_c I} \left[\begin{array}{c} (297 \cdot 7 - 256 \cdot 4)(2 \cdot 15) + 4(347 \cdot 8 - 299 \cdot 5)(2 \cdot 95) \\ + (364 \cdot 5 - 313 \cdot 9)(3 \cdot 75) \end{array} \right]$$

$$= \frac{1186}{32 \times 10^6 \times 6 \cdot 041 \times 10^{-3}}$$

$$= 0 \cdot 0061 \, m$$

Fig. 2·6

Ref.	Example 13.2 Class 2 Member	13.2/9
	Total long-term deflection $\quad= 0.0061 + 0.0061$ $\quad= 0.0122\ m$	$\delta_t = \dfrac{L}{1230}$

Ref.	Example 13.3 Class 3 Member	13.3/1

Pretensioned floor beams are required to span 5m, with the general arrangement shown below.

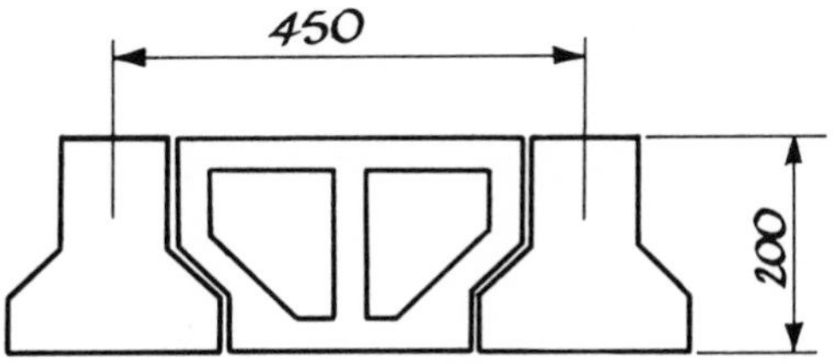

450

200

		kN/m²
Loading	R.C. hollow-core slabs	2·3
	Finishes	1·2
	Partitions	1·0
	Services and Ceiling	0·3
	Imposed Load	3·0

Material Properties

Concrete : f_{cu} at 28 days = 40 N/mm²

f_{ci} at 7 days = 30 N/mm²

Steel : f_{pu} = 1770 N/mm²

f_y = 460 N/mm²

Allowable Stresses

f'_{max} = 15·0 N/mm² f_{max} = 13·3 N/mm²

f'_{min} = -2·5 N/mm²

Ultimate Strength (Preliminary)

Beam self-weight = 0·66 kN/m

Total DL = 0·66 + 0·45 x 2·5 + 0·35 x 2·3

= 2·59 kN/m

Ref.	Example 13.3 Class 3 Member	13.3/2

Ult. uniform load $= 1.4 \times 2.59 + 1.6 \times 0.45 \times 3.0$

$\qquad\qquad\qquad = 5.79 \ kN/m$

Max. ult. BM $\quad = \dfrac{5.79 \times 5^2}{8} \ = \ 18.1 \ kNm$

Initially assume that the steel has yielded

$d = 200 - (20+10) = 170 \ mm$

For internal equilibrium :

$0.87 \times 1770 \, A_{ps} = 0.45 \times 40 \times 100 \times 0.9x$

$0.87 \times 1770 \, A_{ps} (170 - 0.45x) = 18.1 \times 10^6$

$\qquad \therefore \ x = 85 \ mm \ and \ A_{ps} = 89 \ mm^2$

Assuming $\beta = 0.85$

$\qquad \varepsilon_{pe} = \dfrac{0.85 \times 0.7 \times 1770}{205 \times 10^3} = 0.00514$

$\therefore \ \varepsilon_{pb} = 0.00514 + \dfrac{(170-85) \times 0.0035}{85}$

$\qquad\qquad = 0.00864 \ < \varepsilon_2 = 0.00125$

For $\varepsilon_1 \leqslant \varepsilon_{pb} \leqslant \varepsilon_2$

$\varepsilon_{pb} = 0.00514 + \dfrac{(170-x) \times 0.0035}{x}$

$f_{pb} = 1232 + \dfrac{(\varepsilon_{pb} - 0.00601) \times 308}{0.00649}$

$\qquad = \dfrac{1}{x} \left(1025x + 28238 \right)$

Ref.	*Example 13.3* *Class 3 Member*	13.3/3
	For equilibrium:	

For equilibrium:

$$0.45 \times 40 \times 100 \times 0.9x = A_{ps} \times \frac{1}{x}(1025x + 28238)$$

$$0.45 \times 40 \times 100 \times 0.9x\,(170 - 0.45x) = 18.1 \times 10^6$$

$$\therefore x = 85\,mm \quad and \quad A_{ps} = 101\,mm^2$$

Check: $\varepsilon_{pb} = 0.00864$ (*i.e.* $\varepsilon_1 < \varepsilon_{pb} < \varepsilon_2$)

Min. A_{ps} req'd. = $101\,mm^2$

<u>Section Properties</u>

$$A_c = 2.75 \times 10^4\,mm^2$$
$$Z_t = 7.83 \times 10^5\,mm^3$$
$$Z_b = 10.96 \times 10^5\,mm^3$$
$$I = 91.32 \times 10^6\,mm^4$$
$$\bar{y} = 83.3\,mm$$

<u>Prestress Force</u>

$$W_s = 2.59 + 0.45 \times 3.0 = 3.94\,kN/m$$

$$M_s = \frac{3.94 \times 5^2}{8} = 12.3\,kNm$$

Tables 5.3 & 5.4

$$f_{ht} = 5.0 \times 1.1 = 5.5\,N/mm^2$$

Assuming $\beta = 0.85$ $e = 83 - 30 = 53\,mm$

$$\frac{0.85P_i \times 10^3}{2.75 \times 10^4} + \frac{0.85P_i \times 10^3 \times 53}{10.96 \times 10^5} - \frac{12.3 \times 10^6}{10.96 \times 10^5} = -5.5$$

$$\therefore P_i = 79.5\,kN$$

For 4no. 5mm dia. cold-drawn wires:

Table 2.2

$$P_i = 4 \times 19.6 \times 0.7 \times 1770 \times 10^{-3}$$
$$= 97.1\,kN$$

<u>Estimate of Losses</u>

(i) Elastic shortening

$$f_{pi} = 0.7 \times 1770 = 1239\,N/mm^2$$

Ref.	Example 13.3 Class 3 Member	13.3/4

At midspan, $M_i = \dfrac{0.66 \times 5^2}{8} = 2.1 \, kNm$

$m = \dfrac{205 \times 10^3}{26 \times 10^3} = 7.9$

$r = \left(\dfrac{91.32 \times 10^6}{2.75 \times 10^4} \right)^{1/2} = 58 \, mm$

Eq. 4.4 $f_{co} = \dfrac{1239}{\left[7.9 + \dfrac{2.75 \times 10^4}{78 \left(1 + \dfrac{53^2}{58^2} \right)} \right]} = 6.19 \, N/mm^2$

$\therefore \Delta f_p = 7.9 \times 6.19 = 48.9 \, N/mm^2 \ (3.9\% \, loss)$ $\alpha = 0.96$

(ii) Creep

Concrete stress at level of tendons at midspan

$= 6.19 - \dfrac{2.1 \times 10^6 \times 53}{91.32 \times 10^6} = 4.97 \, N/mm^2$

$\therefore$ Average stress $= \dfrac{1}{2} (6.19 + 4.97) = 5.58 \, N/mm^2$

$\varepsilon_c = \dfrac{1.8}{26 \times 10^3} \times 5.58 = 3.86 \times 10^{-4}$

$\therefore \Delta f_p = 3.86 \times 10^{-4} \times 205 \times 10^3 = 79.2 \, N/mm^2$

(iii) Shrinkage

$\Delta f_p = 300 \times 10^{-6} \times 205 \times 10^3 = 61.5 \, N/mm^2$

(iv) Relaxation

For Class 2 wires,

$\Delta f_p = 1.2 \times 0.0025 \times 1239 = 37.2 \, N/mm^2$

$\therefore$ Total long-term losses $= 177.9 \, N/mm^2$ $\beta = 0.82$
$(14.4\% \, loss)$

Ultimate Strength

Since A_{ps} provided is less than the minimum amount determined previously, extra reinforcement is required.

Ref.	*Example 13.3* *Class 3* *Member*	13.3/5
	Assuming that these bars have yielded, the approx. area of additional steel is :	
	$$A_s \approx (101 - 78) \times \frac{1770}{460} = 89\,mm^2$$	*Try 2T10*
	For equilibrium,	
	$$0.45 \times 40 \times 100 \times 0.9x = \frac{78}{x}(1025x + 28238)$$ $$+ 157 \times 0.87 \times 460$$	
	$$\therefore \quad x = 102\,mm$$	
	$$\therefore \varepsilon_{pb} = \frac{0.82 \times 0.7 \times 1770}{205 \times 10^3} + \frac{(170 - 102) \times 0.0035}{102}$$	
	$$= 0.00729 \quad (i.e.\ \varepsilon_1 < \varepsilon_{pb} < \varepsilon_2)$$	
	$$f_{pb} = 1302\,N/mm^2$$	
	$$\varepsilon_{st} = \frac{(170 - 102) \times 0.0035}{102} = 0.00233$$ $$(> \varepsilon_y = 0.002)$$	
	$$M_u = \left[\begin{array}{l} 78 \times 1302\,(170 - 0.45 \times 102) \\ + 157 \times 0.87 \times 460(170 - 0.45 \times 102) \end{array}\right] \times 10^{-6}$$	*Ultimate strength adequate*
	$$= 20.4\,kNm$$	
	Transfer Stresses	
	$$\alpha P_i = 0.96 \times 97.1 = 93.2\,kN$$	
	$$f_t = \frac{93.2 \times 10^3}{2.75 \times 10^4} - \frac{93.2 \times 10^3 \times 53}{7.83 \times 10^5} + \frac{2.1 \times 10^6}{7.83 \times 10^5}$$	
	$$= 3.39 - 6.31 + 2.68$$	
	$$= -0.24\,N/mm^2 \quad (> f'_{min})$$	
	$$f_b = 3.39 + (6.31 - 2.68) \times \frac{7.83}{10.96}$$	
	$$= 5.98\,N/mm^2 \quad (< f'_{max})$$	
	Service Stresses	
	Cracked section analysis :	

Ref.	Example 13.3　　Class 3 Member	13.3/6

Example 13.3　　Class 3 Member

The stress in the bars is thus $< 150 \, N/mm^2$ and the cracking will not be excessive.

Also, the maximum concrete stress is $< f_{max}$.

Shear Resistance

(i) Uncracked sections

Eq's 8.1÷7.4

$$L_t = \frac{600 \times 5}{(30)^{\frac{1}{2}}} = 548 \, mm \; ; \; f_{prt} = 0.24 \times (30)^{\frac{1}{2}} = 1.31 \, N/mm^2$$

At critical section,

Eq.7.8

$$f_{cpx} = \frac{0.283}{0.548} \left(2 - \frac{0.283}{0.548}\right) \times \frac{0.82 \times 97.1 \times 10^3}{2.75 \times 10^4}$$

$$= 0.72 \, N/mm^2$$

Eq.7.6

$$\therefore V_{co} = 0.67 \times 100 \times 200 \left(1.31^2 + 0.8 \times 0.72 \times 1.31\right)^{\frac{1}{2}} \times 10^{-3}$$

$$= 21.1 \, kN$$

$$V = 5.79(2.5 - 0.183) = 13.4 \, kN$$

(ii) Cracked sections

m	0.5	1.0	1.5	2.0	2.5
M_o	5.9	5.9	5.9	5.9	5.9
M	6.5	11.6	15.2	17.4	18.1
V	11.6	8.7	5.8	2.9	0
V_c	0.60	0.60	0.60	0.60	0.60
V_{cr}	17.5	11.4	9.2	7.9	6.9

Eq.7.13

Stresses at Supports

$$f_t = \frac{0.82 \times 97.1 \times 10^3}{2.75 \times 10^4} - \frac{0.82 \times 97.1 \times 10^3 \times 53}{7.83 \times 10^5}$$

$$= 2.90 - 5.39$$

$$= -2.49 \, N/mm^2$$

Uncracked shear resistance adequate

Cracked shear resistance adequate

200 | 83

Ref.	Example 13.3 Class 3 Member	13.3/7

$f_b = 2.90 + 5.39 \times \dfrac{7.83}{10.96} = 6.75 \text{ N/mm}^2$

If the beam is considered as a Class 2 member, $f_{min} = -2.85 \text{ N/mm}^2$.

∴ The beam is uncracked near the supports and no de-bonding is necessary.

Deflections

(i) Transfer   The section remains uncracked at this stage.

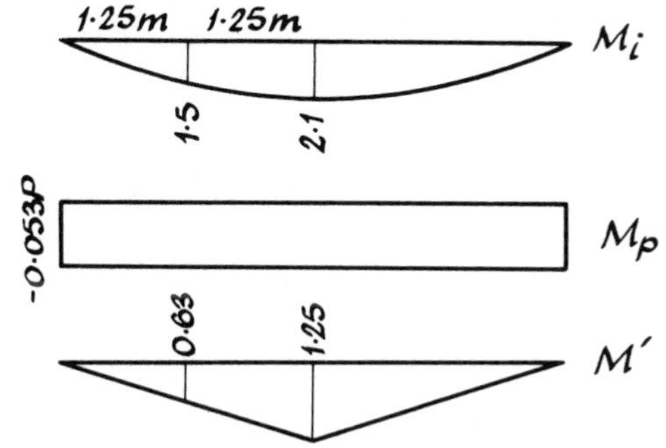

$\delta_i = \dfrac{2 \times 2.5}{6EI} \begin{bmatrix} 4(1.5 - 0.053P)(0.63) \\ + (2.1 - 0.053P)(1.25) \end{bmatrix}$

With $P = \alpha P_i = 0.96 \times 97.1 = 93.2 \text{ kN}$,

$\delta_i = -\dfrac{10.2}{EI} = \dfrac{-10.2}{26 \times 10^6 \times 91.32 \times 10^{-6}}$

$= -0.0043 \text{ m}$

$\delta_i = -\dfrac{L}{1166}$

(ii) Service

In order to simplify the calculations, the beam will be treated as rectangular, with dimensions 200×100 mm

Under service load the beam is cracked, and deflections should be determined assuming that those regions are cracked under permanent and non-permanent loads.

Ref.	Example 13.3 Class 3 Member	13.3/8
	Short-term deflections under permanent load (0.5 Imposed load):	
	Permanent load = $2.59 + 0.5 \times 0.45 \times 3.0$	
	$= 3.27 \ kN/m$	
	$M_{per.} = \dfrac{3.27 \times 5^2}{8} = 10.2 \ kNm$	
	Cracked section analysis gives:	
Table 6.1	$\dfrac{1}{r_c} = \dfrac{0.000418}{145} = 2.88 \times 10^{-6} \ mm^{-1}$	
	$\therefore \delta_a = 0.104 \times 5^2 \times 10^6 \times 2.88 \times 10^{-6}$	
	$= 7.5 \ mm$	
	Short-term deflections under total load:	
	$M_s = 12.3 \ kNm$	
	$\dfrac{1}{r_c} = \dfrac{0.000578}{118} = 4.90 \times 10^{-6} \ mm^{-1}$	
	$\therefore \delta_b = 0.104 \times 5^2 \times 10^6 \times 4.90 \times 10^{-6}$	
	$= 12.7 \ mm$	

Ref.	*Example 13.3* *Class 3 Member*	*13.3/9*

For long-term deflections under permanent load:

$$\frac{1}{r_c} = \frac{0.000409}{152} = 2.69 \times 10^{-6} \ mm^{-1}$$

$$\therefore \delta_c = 0.104 \times 5^2 \times 10^6 \times 2.69 \times 10^{-6}$$

$$= 7.0 \ mm$$

Total deflection $= \delta_b - \delta_a + \delta_c - \delta_i$

$$= 12.7 - 7.5 + 7.0 - 4.3$$

$$= 7.9 \ mm$$

$$\delta_t = \frac{L}{633}$$

BIBLIOGRAPHY

General

Abeles, P. W. and Bardhan-Roy, B. K. (1981) *Prestressed Concrete Designer's Handbook*, Viewpoint, Slough.

Allen, A. H. (1981) *An Introduction to Prestressed Concrete*, Cement and Concrete Association, Slough.

Bate, S. C. C. and Bennett, E. W. (1976) *Design of Prestressed Concrete*, Wiley, New York.

Gerwick, B. C. (1971) *Construction of Prestressed Concrete Structures*, Wiley, New York.

Guyon, Y. (1974) *Limit State Design of Prestressed Concrete*, Applied Science, London.

Krishna Raju, N. (1981) *Prestressed concrete*, Tata McGraw-Hill, New Delhi.

Leonhardt, F. (1964) *Prestressed Concrete: Design and Construction*, Wilhelm Ernst, Berlin.

Libby, J. R. (1977) *Modern Prestressed Concrete*, Van Nostrand Reinhold, New York.

Lin, T. Y. and Burns, N. H. (1981) *Design of Prestressed Concrete Structures*, Wiley, New York.

Naaman, A. E. (1981) *Prestressed Concrete Analysis and Design*, McGraw-Hill, New York.

Nilson, A. H. (1978) *Design of Prestressed Concrete*, Wiley, New York.

Ramaswamy, G. S. (1976) *Modern Prestressed Concrete Design*, Pitman, London.

Rowe, R. E. *et al.* (1987) *Handbook to British Standard BS8110: 1985 Structural Use of Concrete*, Viewpoint, London.

Sawko, F. (Ed.) (1978) *Developments in Prestressed Concrete*, Vols. 1 and 2, Applied Science, London.

Warner, R. F. and Faulkes, K. A. (1979) *Prestressed Concrete*, Pitman, Sydney.

Bridges

Clark, L. A. (1982) *Concrete Bridge Design to BS5400*, Cement and Concrete Association, London.

Green, J. K. (1973) *Detailing for Standard Prestressed Concrete Bridge Beams*, Cement and Concrete Association, London.

Mathivot, J. (1983) *The Cantilever Construction of Prestressed Concrete Bridges*, Cement and Concrete Association, London.

Pennells, E. (1978) *Concrete Bridge Designer's Manual*, Cement and Concrete Association, London.

Podolny, W. (1982) *Construction and Design of Prestressed Concrete Segmental Bridges*, Wiley, New York.

Special topics

American Society of Civil Engineers, *Fatigue Life of Prestressed Concrete Beams*, Bulletin no. 19 of Reinforced Concrete Research Council.

Concrete Society (1977) *Segmental Precast Prestressed Concrete Piles*. Technical Report no. 12.

Cowan, H. J. (1965) *Reinforced Concrete and Prestressed Concrete in Torsion*, London.

Green, J. K. and Perkins, P. H. (1980) *Concrete Liquid Retaining Structures*, Applied Science, London.

Computing

Hulse, R. and Mosley, W. H. (1987) *Prestressed Concrete Design by Computer*, Macmillan, London.

INDEX